T0249645

Enhancing Learning and Teaching through Student Feedback in Engineering

CHANDOS
LEARNING AND TEACHING SERIES

Series Editors: Professor Chenicheri Sid Nair and Dr. Patricie Mertova
(e-mails: sid.nair@uwa.edu.au and patricie.mertova@education.ox.ac.uk)

This series of books is aimed at practitioners in the higher education quality arena. This includes academics, managers and leaders involved in higher education quality, as well those involved in the design and administration of questionnaires, surveys and courses. Designed as a resource to complement the understanding of issues relating to student feedback, books in this series will respond to these issues with practical applications. If you would like a full listing of current and forthcoming titles, please visit our website www.chandospublishing.com or email wp@woodheadpublishing.com or telephone +44 (0) 1223 499140.

New authors: we are always pleased to receive ideas for new titles; if you would like to write a book for Chandos, please contact Dr Glyn Jones on gjones@chandospublishing.com or telephone number +44 (0) 1993 848726.

Bulk orders: some organizations buy a number of copies of our books. If you are interested in doing this, we would be pleased to discuss a discount. Please email wp@woodheadpublishing.com or telephone +44 (0) 1223 499140.

Enhancing Learning and Teaching through Student Feedback in Engineering

EDITED BY
CHENICHERI SID NAIR, ARUN PATIL
AND PATRICIE MERTOVA

CP

CHANDOS
PUBLISHING

Oxford   Cambridge   Philadelphia   New Delhi

Chandos Publishing
Hexagon House
Avenue 4
Station Lane
Witney
Oxford OX28 4BN
UK
Tel: +44 (0) 1993 848726
Email: info@chandospublishing.com
www.chandospublishing.com

Chandos Publishing is an imprint of Woodhead Publishing Limited

Woodhead Publishing Limited
80 High Street
Sawston
Cambridge CB22 3HJ
UK
Tel: +44 (0) 1223 499140
Fax: +44 (0) 1223 832819
www.woodheadpublishing.com

First published in 2012

ISBN: 978-1-84334-645-6 (print)
ISBN: 978-1-78063-299-5 (online)

© The editors and contributors, 2012

British Library Cataloguing-in-Publication Data.
A catalogue record for this book is available from the British Library.

All rights reserved. No part of this publication may be reproduced, stored in or introduced into a retrieval system, or transmitted, in any form, or by any means (electronic, mechanical, photocopying, recording or otherwise) without the prior written permission of the Publishers. This publication may not be lent, resold, hired out or otherwise disposed of by way of trade in any form of binding or cover other than that in which it is published without the prior consent of the Publishers. Any person who does any unauthorised act in relation to this publication may be liable to criminal prosecution and civil claims for damages.

The Publishers make no representation, express or implied, with regard to the accuracy of the information contained in this publication and cannot accept any legal responsibility or liability for any errors or omissions.

The material contained in this publication constitutes general guidelines only and does not represent to be advice on any particular matter. No reader or purchaser should act on the basis of material contained in this publication without first taking professional advice appropriate to their particular circumstances. Any screenshots in this publication are the copyright of the website owner(s), unless indicated otherwise.

Typeset by RefineCatch Limited, Bungay, Suffolk
Printed in the UK and USA.

Contents

List of figures and table

Figures

Table

Preface

Evaluation in higher education aims to verify whether or not an action or a process has achieved the aims and outcomes that were originally envisaged for it. This book is the first in a series on student feedback in specific disciplines and how it has, or can be, used to enhance the quality of teaching and learning in higher education. It follows on from a book entitled *Student Feedback: The Cornerstone to an Effective Quality Assurance System in Higher Education* (Chandos, 2010) which focused on student feedback in higher education in general.

This volume, specifically focusing on student feedback in engineering, will provide insight into how the process of evaluation and the design of evaluation within that discipline are currently utilised in a number of countries around the world. More importantly, the book aims to show the reader that student feedback has a place within engineering and in higher education more generally.

This volume introduces student feedback in engineering and draws upon international perspectives within the higher education setting (Contributions come from Chile, Hong Kong, India, Sweden, Thailand, the UK and Australia). A majority of the contributors are practitioners in the field of engineering and some specialise more generally in student feedback; however, all the chapters show contributors' perspectives on the subject and provide insights into the practices within the contributors' institutions and approaches utilised in their higher education systems.

There are eight chapters in this book. The introductory chapter offers an insight into the fundamentals of student feedback within the discipline. Subsequent chapters delve into the practices, views and approaches to student feedback in higher education systems around the world. The final chapter draws on the information presented in the earlier chapters and outlines the current trends, issues and the future of student feedback in engineering education. Common themes run through the majority of the chapters: the value of utilising student feedback as part of the various quality enhancement approaches within engineering. The main argument of this book is that such feedback is essential in order to improve the key learning outcomes of engineering education, such as enabling skills to understand, communicate and solve problems. It is also clear that, although the use of student feedback is still in its infancy in many parts of the world, it is linked to the realisation that such feedback is critical in enhancing the quality of engineering programmes.

Chenicheri Sid Nair, Arun Patil and Patricie Mertova

About the authors

Anders Ahlberg is an academic developer at the Faculty of Engineering at Lund University (LTH). He is an associate professor in earth science and has worked as an academic developer since 2001. At LTH, he is responsible for the faculty PhD programme and the programmes for doctoral supervisors and associate professors. He is a member of both the Teachers' Academy (Faculty of Natural Sciences) and the Educational Advisory Board (Faculty of Medicine) at Lund University. His recent research interests include strategic educational development in research-dominated academic environments.

Roy Andersson is both an academic developer (50 per cent) and a senior lecturer in computer science (50 per cent) at the Faculty of Engineering, Lund University (LTH). He has been an academic developer (part-time) since 1998 and his main interest is in supporting academics to investigate their practice through research, in other words supporting scholarship of teaching and learning. He is responsible for the programme of teacher training and development at LTH and teaches on the programme himself. He is also a member of the Teachers' Academy at LTH.

Rosario Carrasco graduated in sociology from the Pontifical Catholic University of Chile in 1999, and obtained an MA in higher education from the Ontario Institute for Studies in Higher Education of the University of Toronto, Canada in 2007. She has published articles and book chapters related to accreditation and quality assurance in Chilean engineering programmes. She is currently finalising an externally funded project, which is part of the Ministry of Education's Academic Innovation Fund (MECESUP2). Its objectives included the assessment of the impact of other MECESUP projects implemented at the same university, the design of an online system intended to enhance the supervision of these projects, and the identification of relevant impact indicators.

Ake Chaisawadi has been associate professor in the Faculty of Engineering at King Mongkut's University of Technology Thonburi, Bangkok, Thailand since 1977. His main interest is in measurement, control, instrumentation and digital signal processing. He holds a PhD in system science from Kanazawa University, Japan.

Neelam Chaplot has been in the field of teaching and training since 1996. Currently she is working at Jaipur Engineering College and Research Centre, Jaipur, India. Her main interest is in the fields of artificial intelligence and data mining. She holds an MTech in computer science from Banasthali University, India.

Mousumi Debnath is a professor and head in the Department of Biotechnology at Jaipur Engineering College and Research Center (JECRC), Jaipur, India. Currently she also heads the biotechnology training and research group. She has taught on virtually every course in biotechnology, but has mainly taught genetic engineering. She has published 11 books and 40 articles in peer reviewed journals. She also holds a Master's degree in business administration with specialisation in human resource management. She is an active member of the training and placement group at JECRC. She has also conducted soft skills classes and mock interviews during campus recruitment at JECRC.

Kristina Edström is an educational developer with an engineering background (MSc in engineering, Chalmers University of Technology, Sweden) whose main interest is in driving strategic change processes to improve engineering education, on all levels: course, curriculum, institutional, national and international. She lectures in engineering education development at the Royal Institute of Technology (KTH), Sweden. She leads and participates in educational development activities at KTH, in Sweden and internationally, and serves on the council of the international CDIO Initiative as well as of the European Society for Engineering Education (SEFI). At KTH, over 460 faculty members have taken her 7.5 ECTS credit course *Teaching and Learning in Higher Education*. Kristina received the KTH Education Prize in 2004. As a Chalmers student, she was deeply engaged as a student representative and in 2009, was appointed a lifetime honorary member of the KTH Student Union THS.

Manohar Gottimukkula has been working as a statistician since 2006. Currently, he is a research officer at the National Nutrition Monitoring Bureau, National Institute of Nutrition, Hyderabad, India. He holds a Master's degree in statistics from Osmania University, Hyderabad, India.

His main interests are in academic research, nutrition and community division studies.

Suresh Narain Gupta is an advisor at the Jaipur Engineering College and Research Center (JECRC), Jaipur, India. Currently he also heads the research facilitation unit at JECRC. Prior to joining JECRC, he spent more than three decades at the Indian Institute of Technology, Delhi. His area of specialisation is communication engineering. He has published a large number of papers in various of the *IEEE Transactions* (USA), generally in areas related to signal processing and propagation studies. He leads the training and placement group at JECRC.

Kalayanee Jitgarun has been an associate professor in the Electrical Technology Department, King Mongkut's University of Technology Thonburi, Bangkok, Thailand since 1991. Most of her research work focuses on curriculum development and instruction in vocational education. She supervises students at PhD, Master's and Bachelor's levels. She holds a Doctor of Education in curriculum and instruction from Texas Southern University, USA. She has published widely and has collaborated with colleagues across Asia, Australia and Europe, which has provided her with broad experience in both in education and culture.

David Kane is a senior researcher at the Social Research and Evaluation Unit within the Faculty of Education, Law and Social Sciences at Birmingham City University, Birmingham, UK. David has a keen interest in all areas of higher education policy and practice, particularly the student experience of higher education. He has presented a number of papers at international conferences on this subject and has worked on institutional student satisfaction surveys for a number of higher education institutions since 2005. David is also interested in the application and use of technology, including social media, in learning and teaching and is currently developing this interest with colleagues at Birmingham City University.

Pinit Kumhom has been an assistant professor in the Faculty of Engineering at King Mongkut's University of Technology Thonburi, Bangkok, Thailand since 2001. His main interest is in implementations of DSP algorithms, ASIC/FPGA designs and applications, and embedded system designs. He holds a PhD in electrical and computer engineering from Drexel University, Philadelphia, PA, USA.

Mario Letelier is director of the Centre for Research in Creativity and Higher Education at the University of Santiago of Chile. He is also a member of the National Commission for Accreditation and president of the Chilean Society for Engineering Education. He holds a PhD in mechanical engineering from the University of Toronto, Canada. Since 1983, he has been involved in fostering engineering education in Chile, and participating in the design and implementation of national quality assurance policies. His main area of interest within mechanical engineering is non-Newtonian fluids mechanics.

Daniela Matamala graduated recently in civil industrial engineering at the University of Santiago of Chile and works at the Centre for Research in Creativity and Higher Education at the University of Santiago of Chile. Since 2008, she has been involved in accreditation of undergraduate programmes at that university. Additionally, she has participated in teaching improvement projects developed in conjunction with other Chilean universities, sponsored by the Centre for Inter-University Development.

Patricie Mertova is currently a research fellow in the Department of Education, University of Oxford, England. She was previously a research officer at the University of Queensland, and, prior to that, a research fellow in the Centre for the Advancement of Learning and Teaching and the Centre for Higher Education Quality, Monash University, Australia. She has recently completed her PhD focusing on the academic voice in higher education quality. She has research expertise in the areas of higher education, higher education quality and internationalisation. Her background is also in the areas of linguistics, translation, cross-cultural communication and foreign languages.

Chenicheri Sid Nair is currently at the Centre for Advancement of Teaching and Learning, University of Western Australia (UWA), Perth. Prior to his appointment to UWA, he was a quality adviser (research and evaluation) in the Centre for Higher Education Quality at Monash University, Australia. He has an extensive expertise in the area of quality development and evaluation, and he also has considerable editorial experience. Currently, he is an editor of the *International Journal of Quality Assurance in Engineering and Technology Education*. Prior to this, he was also a managing editor of the *Electronic Journal of Science Education*. He is also an international consultant in a number of countries establishing in quality and evaluations.

Claudia Oliva recently graduated in psychology from the University of Santiago of Chile and since then has been worked at the Centre for Research in Creativity and Higher Education at hat university. She is developing expertise in curricula design, pedagogical innovation and academics training. She has been involved in several projects related to learning assessment, equity analysis and development and coordinates the assessment of new programme proposals for the vice-rector.

Mukeshwar Pandey is presently working in the Department of Biotechnology, Jaipur Engineering College and Research Centre (JECRC), Jaipur. He holds a Master's degree in Biotechnology from Jiwaji University, Gwalior, India and is currently undertaking his PhD there. He has published a number of research and review articles in peer reviewed journals and co-authored a number of book chapters. He has also participated in various national and international conferences. He also participates in campus recruitment at JECRC on a regular basis.

Arun Patil is senior lecturer in engineering at the Central Queensland University, Australia. He has over 18 years of teaching and administrative experience in India and Australia in higher and further education. Arun has published widely, and he is a founder editor-in-chief of the *International Journal of Quality Assurance in Engineering and Technology Education*. With Peter Gray he has edited *Engineering Education Quality Assurance: a Global Perspective* (Springer, 2009). He has organised over a dozen of international conferences in engineering education in various parts of the globe.

Suthee Ploisawaschai is a research fellow at the Learning Institute, King Mongkut's University of Technology Thonburi, Bangkok, Thailand. His interests are in literacy, motivation and learning approaches. He has frequently presented at international conferences and published academic journal papers. He holds a Bachelor's degree in English and philosophy from Khon Kaen University, Thailand and a Master's in educational research from University of Exeter, UK.

Doris Rodés graduated in psychology from the University of La Habana, Cuba. She is affiliated with the Centre for Research in Creativity and Higher Education, University of Santiago of Chile and is currently the coordinator of their accreditation process for undergraduate programmes. She has held various positions, including Deputy Director of the Institutional Support Office, Head of Studies Unit, advisor to the Strategic

Development Direction and organisational analyst at the University of Santiago of Chile. Her professional experience has focused on institutional and undergraduate study programme accreditation, strategic planning and organisational diagnosis. She has written a number of articles on tertiary education.

Torgny Roxå has been an academic developer since 1989. Currently, he is an academic developer at the Faculty of Engineering at Lund University. His main interest is in strategic educational development in higher education, a cultural approach where the essentialist and the socio-cultural perspectives are used in combination. he has developed several major measures for change, both at Lund University and nationally in Sweden. He holds a Master's in higher education from Griffith University, Australia.

María José Sandoval graduated recently in civil industrial engineering from the University of Santiago of Chile and since then has been worked at the Centre for Research in Creativity and Higher Education at that university, She has worked on several projects on institutional development. These include: institutional accreditation, curricular development, institutional capacities and academics training. Her main interest lies in the area of academic innovation and management. She is also currently working on a research project which aims to diagnose the academic profiles of new students.

Kin Wai Michael Siu is a professor in the School of Design and leads the Public Design Lab at Hong Kong Polytechnic University. His main research interests are in design culture and theory, and design and engineering education. He has worked on consultancies in design and engineering programmes. He is the subject specialist for the Hong Kong Council for Accreditation of Academic and Vocational Qualifications. He has published more than 150 refereed papers. He has won numerous international design and invention awards, and owned more than 50 US and Asian design patents.

Pramod Kumar Tiwari is a senior advisor at the Jaipur Engineering College and Research Center Foundation. He is a former Indian Police Service officer. He has had an extensive exposure to policing in Rajasthan, having worked in the field as District Superintendent of Police as well as Deputy Inspector General of Police in Ranges. He was awarded the Police Medal for Meritorious Services in 1989 and the President's Police Medal for Distinguished Services in 2001. He has lectured on various aspects of management at various fora. His area of special interest is in soft skills.

James Williams is a senior researcher at the Centre for Research into Quality, Birmingham City University, Birmingham, UK. James researches aspects of the broad experience of students in higher education, at national and international levels, and has published widely in the field. Much of this is drawn from his experience in collecting and using institutional student feedback. He regularly presents papers and holds workshops at international conferences on this subject and has coordinated several institutional student satisfaction surveys since 2004. He is associate editor of the international journal, *Quality in Higher Education*. A historian by training, he continues to publish within his original field of Tudor cultural history, and is currently researching the history of higher education in the non-university sector.

Student feedback in engineering: a discipline-specific overview and background

Kristina Edström

Abstract: This chapter reviews the need to improve key learning outcomes of engineering education, among them conceptual understanding, solving real problems in context, and enabling skills for engineering such as communication and teamwork. At the same time it is necessary to improve both the attractiveness of engineering to prospective students and retention in engineering programmes. Research suggests that to address these problems the full student learning experience needs to better affirm students' identity formation. Student feedback is identified as a key source of intelligence to inform curriculum and course development. An argument is made for clarifying the purpose of any student feedback system, as there is an inherent tension between utilising it for accountability or for enhancement. An example shows how enhancement is best supported by a rich qualitative investigation of how the learning experience is perceived by the learner. Further, a tension between student satisfaction and quality learning is identified, suggesting that to usefully inform improvement, feedback must always be interpreted using theory on teaching and learning. Finally, a few examples are provided to show various ways to collect, interpret and use student feedback.

Key words: problem solving, engineering skills, curriculum development, learning experience, identity, investigation.

Introduction

The aim of this book is to provide inspiration for *enhancement of engineering education* using student feedback as a means. It is important to recognise that enhancement is a value-laden term, and the course we set must be the result of the legitimate claims of all stakeholders, among them students, society, employers and faculty. As external stakeholders society and employers are mainly interested in the *outcomes* of engineering education – such as the competences and characteristics of graduates, the supply of graduates and the cost-effectiveness of education. Students and faculty share an additional interest in the teaching and learning *processes*, as internal stakeholders. Here, student feedback will be discussed as a source of information which can be productively used to improve engineering education – both the outcomes and processes – in the interest of all the stakeholders. As will be shown, the focus on enhancement has far-reaching implications for shaping an approach to collecting, interpreting and utilising student feedback: the keyword is usefulness to inform improvement.

Improving the outcomes of engineering education

The desired outcomes of engineering education are shown in Figure 1.1, categorised into several layers. Each of these aspects can be the focus for improvement, and thus the underlying rationale for engineering education enhancement. Here the layers are nested in the sense that quality at any of the levels depends on the lower levels, and interrelations between the levels are crucial. An intervention to improve any one of the aspects needs to be seen in this full context. Ultimately, any outcomes of engineering education should be discussed in relation to the highest aim, which is to produce graduate engineers capable of purposeful professional practice in society.

The target for improvement can range from a detail, such as students' conceptual understanding of a single concept in the subject, to a much more complex outcome such as their overall ability to contribute to a sustainable society. Nevertheless, whatever aspect one wishes to enhance, it is always within the context of the full curriculum, and singling out one aspect and addressing it with an isolated intervention will therefore only have a limited impact. For instance, if the aim is to develop graduates who would be more innovative engineers, it is not enough to insert a

Figure 1.1   Desired outcomes of engineering education

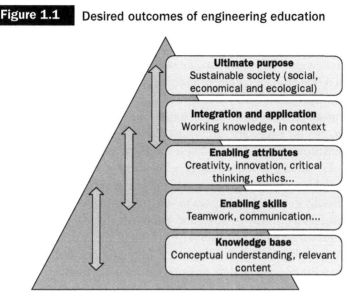

single 'innovation learning activity', such as asking the students to brainstorm 50 ways to use a brick. This is indeed an enjoyable activity, but as a bolt-on intervention it will have a limited impact and fail to truly foster innovative engineers. A successful endeavour must address relevant aspects of the whole curriculum: selection of content, required conceptual understanding, integration and application of knowledge, and the enabling skills and attributes needed for innovation. At the same time, innovation must be seen in a purposeful context – innovation for what and for whom?

Improving problem-solving skills

The rhetoric in engineering education is that engineers are problem solvers, and therefore much effort in education is devoted to developing students' proficiency in problem solving. In lectures, tutorials and textbooks, students encounter numerous problems, and they quickly develop the habit of plugging numbers into equations and arriving at a correct answer, remarkably often a neat expression like $\pi/2$. However, it is not unusual to discover that a significant proportion of students, even many of those who have successfully passed an exam, display poor understanding whenever they are required to do anything outside reproducing known manipulations to known types of problems.

Educators are often surprised and disappointed by this, because the intention was that students should be able to explain matters in their own words, interpret results, integrate knowledge from different courses and apply it to new problems. In short, an important outcome of engineering education is that students acquire the conceptual understanding necessary for problem solving.

But if problem-solving skills are important outcomes of engineering education, it is necessary to widen the understanding of what should constitute the *problems*. While students indeed encounter many problems in courses, an overwhelming majority are pure and clear-cut, with one right answer: they are textbook problems. In fact, they have been artificially created by teachers to *illustrate* a single aspect of theory in a course. Thus, much of students' knowledge can be accessed only when the problems look very similar to those in the textbook or exam: what they have learned often seems inert in relation to real life. This is because real life consists mainly of situations that are markedly different from what students are drilled to handle. Real-life problems can be complex and ill-defined and contain contradictions. Interpretations, estimations and approximations are necessary, and therefore 'one right answer' cases are exceptions. Solving real problems often requires a systems view. In the typical engineering curriculum students seldom practice how to identify and formulate problems themselves, and they rarely practice and test their own judgement. Students are simply not comfortable in translating between physical reality and models, and in understanding the implications of manipulations. As engineering is fundamentally based on this relationship, it certainly seems as if engineering educators have some problems to solve in engineering education.

Real problems also have the troublesome character that they do not fit the structure of engineering programmes. Real problems cannot easily fit into any of the subjects because they do not have the courtesy to respect (the socially constructed) disciplinary boundaries. As both modules and faculty are organised into disciplinary silos, at least in the research-intensive universities, real problems seem to be outside everyone's responsibility and the consequence is that students very rarely meet any problem that goes across disciplines. To make matters worse, real problems are not only cross-disciplinary within engineering, but as they are often rich with context factors such as understanding user needs, societal, environmental and business aspects, they cross over into subjects outside engineering. While many students' first response is to define away all factors that are not purely technical and then give the remainder of the problem a purely technical solution, this solution is probably not adequate to address the original problem. Engineers also need to be able to address problems that have a real context.

If this list of shortcomings seems overwhelming, the good news is that the situation certainly can be improved, as the fault lies primarily in how engineering is taught and – not least – what is assessed and how, because 'what we assess is what we get'. By constantly rewarding students for merely reproducing known solutions to recurring standard problems in exams, this is what engineering education is reduced to. Enabling students to achieve better and more worthwhile learning outcomes is necessary. Rethinking the design of programmes and courses also makes it possible.

Improving the enabling skills and attributes for engineering

While conceptual understanding and individual problem-solving skills are necessary outcomes of engineering education, they are not sufficient. As engineers, graduates must also be able to apply their understanding and problem-solving skills in a professional context. Because the aim is to prepare students for engineering, education should be better aligned with the actual modes of professional practice (Crawley et al., 2007). This means that students need to develop the enabling skills for engineering, such as communication and teamwork, and attributes such as creative and critical thinking.

It is important to think of these skills and attributes not as 'soft' additions to the otherwise 'hard core' technical knowledge of engineering, but as engineering skills and legitimate outcomes of engineering education. For instance, the ability to communicate in engineering can be expressed in many different ways. Some signs that can be seen as indicators that students have acquired communication skills are when they can:

- use the technical concepts comfortably;
- bring up what is relevant to the situation;
- argue for or against concepts and solutions;
- develop ideas through discussion;
- present ideas, arguments and solutions clearly in speech, in writing, in sketches and other graphic representations;

- explain the technical matters for different audiences;
- show confidence in expressing themselves within the field.

When observing these kinds of learning outcomes in students, it should be impossible to distinguish the communication skills from their expression of technical knowledge, as these are integrated. In short, students' technical understanding is *transformed* by the enabling skill into working knowledge (Barrie, 2004). There is nothing 'soft' or easy about engineering communication skills. In fact, they should be regarded as a whole range of necessary engineering competences embedded in, and inseparable from, students' application of technical knowledge. The same kind of reasoning can be applied to other enabling skills such as teamwork, ethics, critical thinking, etc. A consequence is that these competences should be learned and assessed in the technical context supported by engineering faculty. They cannot be taught in separate classes by someone else. A synergy effect is that the learning activities where the students practice communication in the subject will simultaneously help reinforcing their technical understanding. Thus, developing these enabling skills and attributes is fundamentally about students becoming engineers.

Understanding and improving the full student learning experience

Improving the outcomes of engineering education and the effectiveness of its teaching and learning processes could be characterised as doing what educators already set out to do, only better. That does not imply that it is an easy task, or that it is trivial. It is an ambitious undertaking, an endeavour well worth spending all efforts on. Investigations of teaching and learning processes will help identify discrepancies between educational intentions and what actually happens: such gaps indicate room for improvement. How the learning experience is *perceived by the learner* is probably the best source of intelligence on how education can be improved and it is therefore crucial to seek and interpret student feedback on their concrete experiences. Student feedback will help inform the development of teaching and learning in order to more effectively contribute to the intended educational gains. This is not at all, as will be demonstrated later in this chapter, the same as giving the students what they want: their feedback must always be interpreted.

However, improving engineering education within the framework of present thinking is not enough. The effectiveness in fulfilling society's need of engineering graduates is seriously reduced by two problems. The first problem is the weak recruitment of students to engineering programmes. At least in the industrialised world, engineering education faces considerable problems with attractiveness to prospective students, in general but also particularly in relation to gender. Engineering, as an education and career, is perceived not to accommodate personal development and being passionate about one's work. The ROSE study (Schreiner and Sjøberg, 2007) shows that while secondary school students rate it highly important to work with 'something I find important and meaningful' (females more than males) and definitely agree that science and technology are 'important for society', they give very low ratings to 'I would like to get a job in technology' (females less than males). In less developed countries the picture is somewhat different, as to a greater extent, young people associate engineering with growth and building the country. There is increasing evidence that the problem lies in the experience of being an engineering student in relation to their personal identity formation. Interviews with 134 students (Holmegaard et al., 2010) showed that students who choose engineering are interested in doing engineering themselves: solving real problems in an innovative and creative atmosphere. What they encounter when starting engineering education is often not what they sought, but instead courses focused on textbook examples in disciplinary silos, with very little project-based cross-disciplinary real-world innovative work. Holmegaard et al. (2010) put it bluntly:

> The conclusion is that engineering to a large extent matches the expectations of those who do not choose engineering.

In fact, engineering programs can be so much built from the bottom up that it literally takes years to reach the courses which could remind the students why they wanted to study engineering in the first place. But by that time, many of them will have left the programme.

The second problem is the high drop-out rate in engineering programmes. This is often believed to be an inevitable, almost desirable quality, a weeding out process where the less able students drop out, while the more able persist. But in fact, there is not much difference between those who leave and those who stay, neither in terms of interest in the subject nor in their ability to do well in the courses. The students who persist report having the same problems and disappointments with the curriculum that made their friends leave, and thus the students who leave are only the visible tip of the iceberg (Seymour and Hewitt, 1997). Recent research shows (Ulriksen et al., 2010) that a predictor of persistence is a *sense of belonging*, the degree to which students' identity formation finds a match between who they are, or want to be, with the educational and social experience in the programme. It is as if students have a sensor constantly probing 'Is this me?' For prospective students, if the answer is 'no', they will probably not consider engineering education. If they choose engineering, and the answer during the program turns into a 'no', they are at risk of leaving the programme.

The two problems, lack of attractiveness and low retention, are two sides of the same coin: they constitute the external and internal symptoms of the same identity mismatch. In order to attract students to engineering education, and to retain them until they have successfully completed their degrees, the full learning experience in engineering curricula must support student learning and personal development through providing a meaningful and motivational context. It is hard to see a way to meaningfully tackle this task without the use of student feedback. Because there is a need to understand the relationship between student identity work and the design of engineering programmes, courses and learning activities, student feedback is the best source there is to truly understand how to educate engineers in a way which is aligned with student motivation. It is necessary to adopt an open, sensitive and listening mindset, something of an ethnographic approach to studying and understanding the student life world. Again, this is not the same as giving the students exactly what they say that they want. To address the double-sided problem of attractiveness and persistence, educators must better understand how the full educational experience can match students' identity work so that the reading on 'Is this me?' can be and remain positive. This creates a need to consider not only how to improve the outcomes of education, but this task has to be approached in a way that simultaneously addresses even more fundamentally the student learning experience. 'More of the same' will not tackle the problem – rethinking and action is needed, in an iterative development process.

The task for educators is to engage students in learning experiences which support them in achieving the intended outcomes, as discussed above, and at the same time make the full learning experience affirm their identity formation. Improving the outcomes of engineering education and improving the full student learning experience should not be seen as two separate tasks, because fortunately the two issues are related in such way that they have to some extent the same solutions. The overemphasis on reproducing solutions to artificial and de-contextualised problems is just as inadequate in relation to student motivation as it is to their understanding. Students will more easily find meaning, motivation and personal development in learning experiences which result in conceptual understanding, in developing engineering skills and attributes, in working with real problems in context, in aligning education with professional practice, and in a purposeful approach to engineering in society.

Clarifying the purpose of collecting student feedback

The two classic purposes of evaluation are quality *assurance* and quality *enhancement* (Biggs and Tang, 2007). Other terms used for the same dichotomy are accountability and improvement (Bowden and Marton, 1998); appraisal and developmental purpose (Kember et al., 2002); and judgemental and developmental purpose (Hounsell, 2003). It is important to decide on the purpose from the outset, for two important reasons which will be outlined below. As the title of this book suggests, the focus here is *enhancement* of engineering education. Unfortunately, it is very often the case that the purpose for which the student feedback is being collected is never made clear, and this confusion propagates to every issue surrounding student feedback, including how to collect, interpret and use it. Here an attempt will be made to sort out the different possible purposes and their implications, in order to adopt a truly enhancement-led approach.

Accountability or improvement: a fundamental tension

The first reason for settling the purpose is that there is a tension between accountability and improvement: there is a different logic to them. This can shed some light on why student feedback is often a very sensitive issue. When the purpose is not made clear and allowed to guide every aspect of the system, this confusion can lead to a conflict between stakeholders, a conflict which most often appears around methods for collection and use of student feedback. In fact, student feedback is an issue full of power and politics, and different stakeholders manoeuvre to further their own position around the issue. The cause of this are changed forms of management in the public sector in the last few decades, not only in higher education. Management takes place less through planning processes, rules and regulations, and more through objectives and follow-up (Power, 1999; Dahler-Larsen, 2005). The result is an increasing pressure for evaluations and transparency, and ever more, and ever more sophisticated, systems for evaluation.

The pressure to evaluate and collect student feedback seems to mostly concern *that it is done*. There is considerably less pressure to show any real results from it. But it is necessary to start by clarifying the purpose and utilisation:

> In principle, evaluation should not be made at all unless those making or requiring the evaluation are sure how they are going to benefit from it. (Kogan, 1990)

Evaluation is often loosely coupled (Weick, 1976) to improvements. It is not unusual that the official discourse around a student feedback system states that the purpose is to improve education, while in reality it is being used to audit teachers, or, indeed, only as a ritual intended to create a facade of rationality and accountability (Dahler-Larsen, 2005; Edström, 2008). Evaluation is demanded, despite its lack of actual effects, but mainly because it seems appropriate, and especially so when it involves student feedback. The suggestion here is not to do away with evaluation and student feedback, but that evaluation and collection of student feedback will potentially have much better impact on enhancement of teaching and learning if its purpose is made clear. A purely enhancement-led approach to student feedback can liberate teachers from feeling that they themselves are under scrutiny, and may therefore better lead to improvement.

Here it may be worth noting that accountability is also a legitimate purpose. This is not an argument for a system where educators should not be held accountable for the quality of their work. Educators should be accountable. *But so should evaluators* or anyone arguing for evaluation. If the espoused purpose is enhancement, the system must be designed, all the way through, so that it can indeed serve this purpose. One fundamental problem is that it is difficult to bring about enhancement to teaching and learning through an evaluation system which is accountability led, as it can be perceived as a threat to the teachers, who may react by watching their backs and trying to gloss over any problems. It takes a non-threatening enhancement-led system to liberate teachers and open up for a genuine focus on improvement of education. A survey system alone will not improve education (Kember et al., 2002; Edström, 2008) – but student feedback *can* lead to improvement of education if data is purposefully collected, interpreted and analysed, and turned into action plans coupled with resources, support and leadership.

Accountability or improvement – implications for methodology

The second reason why it is important to decide from the beginning whether the purpose is improvement or accountability, is that different data collection methods would be appropriate depending on which purpose it is supposed to serve. Indeed, it is very common in universities to have heated debates on the technicalities in evaluation methods (for instance items in a survey instrument) without even having discussed the purpose of evaluation and how it is to be utilised. There is often reason to note (Ramsden, 1992) that:

> . . . collecting data is not the same as improving or judging teaching.

Most survey instruments used for collecting student feedback are based on quantitative ratings (Richardson, 2005) and therefore seem designed with an accountability purpose in mind, or at least, they have very limited value to inform improvement. If the purpose is improvement, a qualitative, 'open' approach is better suited to investigate student experience of the teaching and learning processes. With richer descriptions, the analysis of what is going on will be better informed, as will the design of interventions. Table 1.1 illustrates two different ways to ask questions, with completely different potential for analysis and action.

Table 1.1   Two ways of asking for feedback (adapted from Handal, 1996)

Method	Rating (quantitative)	Forward-looking feedback (qualitative)
Question	The laboratory exercise booklet supported my learning: ■ strongly disagree ■ disagree ■ neither agree nor disagree ■ agree ■ strongly agree	Please share your views on the laboratory exercise booklet, in particular how it could be improved:
Responses	63% agree and strongly agree 25% neither agree nor disagree 12% disagree and strongly disagree	■ clearer instructions (>40) ■ basically good (>30) ■ too short notice for preparation task (>20) ■ comments on spelling, etc. (>15) ■ tell us what to do step-by-step (>10) ■ misunderstanding in first measurements (>10) ■ difficult or unnecessary theory (>10) ■ theory part not enough to understand the lab (<10) ■ just confusing (~5) ■ Figure 2 difficult to interpret (~5) ■ theory in the book is better, just refer us there (1) ■ add table of contents (1)

Analysis and actions	63% positive ratings should mean that most students are satisfied. The ratings contain no information that could inform action.	Several suggestions can go straight into the action list: add a table of contents, fix spelling, re-scan Figure 2, clear up an unfortunate misunderstanding regarding the first measurement, and include the booklet in the package handed out on the first day of the course. Students who ask for clearer instructions or even 'cookbook style' are probably accustomed to that from previous courses and seem challenged here by the requirement to create the setup. This may be a sign of a surface approach to learning. They could also have been poorly prepared, to which the late distribution of booklets certainly contributed. In class, confront student expectations of a recipe to follow, by discussing the purpose of the laboratory exercise and how the challenge for more independent laboratory work will require better theoretical preparation but also contribute to quality learning. At the next opportunity, ask for student comments about learning from the exercise ('What would you like to say to next year's students about challenges they may encounter and what they can learn from the exercise?'). Then show a couple of insightful comments to the next cohort of students. The feedback relating to theory is puzzling and needs more investigation. Wait and see if some of the confusing concerns about theory and its relation to this laboratory exercise are reduced by confronting the cookbook expectations. There may also be more behind this: keep worrying and keep eyes and ears open in order to find out more.
Comment	Since the ratings cannot directly inform development, the argument is that a rating system still contributes to enhancement by enabling comparisons across courses or across time. But it is dubious whether ratings can be meaningfully compared across contexts, as data is so aggregated. At most it can serve as a smoke detector, where low ratings may trigger further investigation and analysis.	Consider asking this question in a sheet to be handed in straight after the laboratory class, while students have it fresh in their minds, rather than at the end of the course. As the purpose is purely improvement, the tedious counting of answers is mostly unnecessary. A single student's suggestion may be immediately useful, while another suggestion, even if shared by many students, may be caused by a surface approach to learning and should thus not be accommodated. In fact, it could be directly counterproductive from a learning perspective to satisfy them, but it is still valuable to know that these views exist: it can help design other interventions aimed at better supporting these students.

Using student feedback to improve engineering education

Do not give the students what they want – give them something better!

So how is it possible to go beyond simply collecting data – and use student feedback to actually improve courses and programmes? Interpretation of student comments, and subsequent course development, must be done with knowledge of the context and theories on student learning. In particular, it is important to note that the purpose of higher education is learning, not student satisfaction. This is not to say that an educator's job is to keep students unhappy either, but the relation between learning and satisfaction is not straightforward and there is definitely a tension between them. This is aptly put by Gibbs (2010):

> What [students] may want teachers to do may be known from research evidence to be unlikely to result in educational gains.

Take for instance this e-mail, which a professor received from a student in the first-year mechanics course:

> After looking through all the exams written by you, it feels like you have a much higher standard than the other examiners. The biggest problem, as I see it, is that I cannot learn all your questions because they don't seem to recur. When the questions are always different it is very difficult to learn all the solutions. Please could you make an exam for next week more like those by the other examiners.

Of course, this professor wants students to achieve conceptual understanding, and therefore avoids assessing the most obvious reproduction of standard solutions to known problems. It would be devastating to accommodate this student's wish for more predictable assessment which would allow for, and even invite, learning by rote. But knowing that this attitude exists is still very valuable information, because it can then be addressed in many ways. While there may be limited room to effectively help this particular student in the remaining week before the exam, it is certainly possible to support future students in becoming familiar with the required level, and with this way of expressing understanding, early on in the course. The professor can discuss different qualities of understanding and make it clear what they will set out to achieve in this course, and give little quizzes regularly during the course so it will not come as a shock at the end, when it is too late. Further, the e-mail can spark discussions about the quality of learning outcomes in the department, because the other examiners would probably benefit from a more sophisticated understanding of the relationship between how assessment is designed and its effects on learning.

In relation to this example, a final observation can be made that this professor obviously comes across as rather odd to this student, who, if asked, may give the professor a low rating. Some institutions have student rating system where teachers are rewarded for high ratings, or even depend on them for tenure. That creates incentives to do what it takes to keep students happy, even when it means sacrificing the quality of learning outcomes. But a better principle is: *Do not give the students what they want – give them something better!* Students are learners, not consumers. If what really matters most is the quality of learning and improvement of learning processes, then student feedback must be a much broader concept than student ratings of teaching. But educators must learn how to use student feedback productively; to become useful it must be interpreted:

> To be effective in quality improvement, data [. . .] must be transformed into information that can be used within an institution to effect change. (Harvey, 2003)

A framework for interpreting student feedback

Student feedback shows how the learning process is experienced by the learner, and the main reason to collect their feedback is to be able to improve the effectiveness of course design. Naturally, as was argued above, students will always give their feedback from their own frames of reference. Students' experience of their learning processes – and thus the feedback they give – depends to a great extent on how the learning activity, course or programme are designed. In particular, it will influence the extent to which students adopt a deep or surface approach to learning (Marton and Säljö, 1984; Gibbs, 1992; Biggs and Tang, 2007).

A deep approach to learning is when the student's intention is to find meaning, to understand: the result is well-structured and lasting knowledge. Course characteristics associated with a deep approach are (Biggs and Tang, 207):

- motivational context;
- learner activity, including interaction with others;
- well-structured knowledge base;
- self-monitoring (including awareness of one's own learning processes, self-assessment and reflection exercises).

These should be guiding principles for enhancing engineering education. On the other hand, a surface approach to learning is associated with an intention to complete the task as quickly as possible. The focus is on being able to produce the required signs of knowledge, not the underlying meaning. In anticipation of the assessment, students focus on being able to reproduce the subject matter – but since they do not seek meaning, they will not find it. It is inefficient to study using a surface approach, because the result is disastrous. Although the focus was on passing the course, the resulting learning is poorly structured and easily forgotten, so in fact, passing the course may sometimes be hard – unless assessment allows for this poor quality of understanding. Course characteristics associated with a surface approach are (Gibbs, 1992):

- heavy workload;
- relatively high class contact hours;
- excessive amount of course material;
- lack of opportunity to pursue subjects in depth;
- lack of choice over subjects and lack of choice over the method of study;
- a threatening and anxiety-provoking assessment system.

Many engineering programmes suffer from several of these characteristics and these ought to be reduced.

Most students have the capacity to adopt both approaches, and the purpose of course design is thus to influence students to adopt a deep approach to learning, as the following quote from a third-year engineering student indicates:

> The things I remember from a course are the parts we had assignments on. Then I really sat down with the problems and worked out the solutions myself. When I study old exams, I check up the correct answer right away, and then move on without really learning. (Edström et al., 2003)

This shows that the same student can use both deep and surface approaches, within the same course. Here, working on assignments is associated with a deep approach, while studying for the exam is associated with a surface approach.

In addition to course-related factors, student expectations will be shaped also by other factors preceding the particular course: students' previous experiences and their conceptions of learning (Prosser and Trigwell, 1999), as well as their attitude to knowledge (Perry, 1970). These factors are therefore also present in their learning experience and thus in their feedback. While these factors may predispose a student to spontaneously adopt a deep or surface approach to learning, it is not a fixed and inherent trait but one that can be influenced. In order to support students in assuming a role as learners that differs from their expectations, faculty can actively support their transition into a more appropriate role, by increasing students' awareness of their own learning processes, by confronting the symptoms of the surface approach, and creating trust in the learning model.

Collecting student feedback – a student learning focus

When setting out to collect student feedback, a useful principle is to 'begin with the end in mind'. What data can actually inform education enhancement, or in other words, what is necessary to investigate about the learning process in order to be able to develop it? Educators' own views on teaching and learning will influence which variables are perceived as possible, acceptable and desirable to manipulate in course development. Perhaps this is a contributing reason why so often a focus on teaching and the teacher seems to be present when student feedback is solicited. But if the aim is to improve learning, the guiding principle for the investigation should be 'How are the students doing?' rather than 'How is the teacher doing?' A few suggestions intended to illustrate some of the implications for student feedback that follow from adopting a genuine student learning focus are given below.

Assessment data as student feedback on their learning

If the aim is to improve student learning, then an appraisal of *actual* student learning outcomes is probably the best input to inform course development. To say the least, how well students learned must always be a more relevant variable to pursue than how well they liked the teaching or the teachers. While it is seldom thought of as such, one excellent form of student feedback on their learning is assessment data. Even in a seemingly well-designed and popular course, all teachers who go through assessments will be able to set urgent and relevant improvement targets. And since teachers have to go through assessments anyway, why not make full use of this source of intelligence? A simple approach is to keep a notepad handy when marking student work, to jot down troublesome issues, analysis and possible actions. An added advantage is that the assessments are conveniently available. The data has already been collected and has probably been archived over several years. But the quality of assessment data still depends on the validity of the assessment: does it really measure the intended learning outcomes? Do the exams require the level of conceptual understanding that students were intended to reach, or could it be possible to do well on the exams by only applying lower level outcomes, such as pattern matching the recurring types of problems? Sometimes it may be illuminating to triangulate, for instance by oral discussions that more easily reveal those misconceptions which can hide behind mere reproduction.

Student feedback as a way of directly improving learning

Soliciting feedback can serve as a means to facilitate student awareness of their own understanding and learning processes. In this case, the chief aim is to directly improve learning through reflection. Two examples are given below.

Eric Mazur (1997) developed a method where student feedback is used to expose student understanding as it develops in real time and to immediately improve it. In a lecture, students are given a multiple-choice question where wrong alternatives reflect common misconceptions. After a first vote (using show-of-hands, coloured slips or electronic response systems), students are asked to convince their neighbour for a few minutes, and thereafter the class takes a new vote. Naturally, the discussion is a very effective learning activity for the students, as they will investigate the different assumptions and compare the arguments for and against them, and the class systematically converges towards the correct answer.

The one-minute paper (Angelo and Cross, 1993) is a simple method for soliciting feedback. At the end of any learning activity, students answer one or more questions on a sheet of paper. Questions such as 'What are the three most useful things you have learned?' can be asked, either in relation to the learning activity or to the whole course so far. This will invite student reflection as well as expose to the teacher how the students are getting along, thereby showing the potential for improvement. A common variant is to ask about the muddiest point in a lecture: 'What point remains least clear to you?' How this feedback is used depends on the purpose and the available resources. Selected muddy points can be commented on in the next lecture, via e-mail or the course homepage. Alternatively, the sheets can be stored and only read when preparing for the same lecture next year.

Finding out what students do out of class

While teachers often solicit student feedback on what they, the teachers, do during class, it is much less common to find out what the students themselves do outside scheduled hours. This negligence is remarkable, since the volume and quality of independent study is of utmost importance for the quality of student learning. It is therefore useful to know how much time they spend studying (time-on-task), the distribution of their effort during the course, and whether the teaching generates appropriate learning activity outside class (Gibbs, 1999). Even if it comes as an unpleasant surprise, discovering that most students have not yet opened the course book five weeks into the course is like receiving a precious gift – low hanging fruit – this situation has much potential for major improvements. Finding out can be as simple as asking students to draw a graph indicating how much time they have spent studying every week of the course, or just an open-text question where they are asked to say how they went about their studies in the course. To gain deeper insights, it is a useful to conduct in-depth interviews with students, to investigate their lived world as students. By recording and perhaps even transcribing the interviews their narrative can be analysed just like any research material.

A case of effective learning – finding out why it worked

An innovative learning activity used at the Royal Institute of Technology in Stockholm is student problem-solving sessions replacing the traditional weekly tutorials where a teacher solves problems on the board. Instead, the students will all prepare to present solutions to weekly problem sets. Arriving at the session, students tick on a list what problems they are prepared to present. For each problem, a new student is randomly picked to present at the board, followed by class discussions on any alternative solutions, critical aspects and inherent difficulties. Ticking (say) 75 per cent of the problems is required, but as the purpose is purely formative, the quality of the presentation does not affect the grade. Students must, however, demonstrate that they have prepared, and at least be able to lead a classroom discussion toward a satisfactory treatment of the problem. The teachers who first began to implement this were surprised by the extraordinary results. There were substantial improvements in understanding and in exam results (pass rate going up from below 60 per cent to consistently around 70–85 per cent). The activity was popular with the students (rated well over 4 on a Likert scale of 1–5). Further, teachers appreciated the improved cost-effectiveness (not least by eliminating the tedious work of correcting poor exam papers) and they found themselves in a much more stimulating role, as discussing the subject with students who were prepared and active was much more enjoyable than presenting solutions to a silent room of students taking notes.

But it was student feedback which helped the teachers finding out *exactly why* this process was working so well. Student revealed in interviews that this felt like an effective way to study, not least because the problem-solving was aligned with the performance expected in the assessment. Students actually spent 6–7 hours in preparation before each session, forming groups and running practice sessions in empty classrooms, taking turns to present and discussing each problem together. Student problem-solving sessions were generating much time spent on-task, which was well distributed over the length of the course, and the learning activity was perfectly appropriate (Gibbs, 1999). In fact, the scheduled sessions were just the tip of the iceberg that drove large volumes of extremely good studies. In stark contrast, when asked about preparations for traditional teacher-led sessions, a typical student reply was:

> They are normal tutorials where *he* solves problems, right? No, I don't prepare for that.

At the sessions, all students could easily follow the solution, as they were already familiar with the problem, even if they had not succeeded in solving it. They also received feedback on their own efforts through the class discussion on alternative solutions and critical aspects of the problem. It may not seem like a big difference to a person who accidentally opens the door of a classroom if it is a student or teacher who is standing at the board, but student feedback revealed the difference in volume and quality of work that these learning activities generate outside class. It turned out that student-led and teacher-led tutorials are worlds apart: the learning process is *fundamentally changed*. Student problem-solving sessions obviously improve students' conceptual understanding of the subject. The activity format also contributes to developing communication skills, which is an important enabling skill for an engineer.

Furthermore, it is an active and more social learning format in which both engineering students and teachers enjoy much more stimulating roles than in the old and weary routine. The evidence gained through student feedback played an important role in understanding exactly how the learning process was changed by this particular intervention. With this knowledge and evidence it is possible to spread, if not the exact format of this particular activity but rather its fundamental principles, to benefit other courses in engineering education.

Conclusions

This chapter has argued that student feedback is a crucial source of intelligence which reveals clues about the inner workings of the teaching and learning processes, thus enabling educators to better understand and improve them. Engineering education needs to enhance not only the quality of learning outcomes, but also the full learning experience in order to address the issues of attractiveness and retention in engineering education. An enhancement-led approach to collecting student feedback should be non-threatening to the teaching staff and collect rich qualitative descriptions that can be utilised to inform development. Finally, as the tension between student satisfaction and learning is recognised, student feedback always must be interpreted. The principle is: Do not give students what they want, give them something better!

References

Angelo, T.A. and Cross, P.K. (1993) *Classroom Assessment Techniques*. San Francisco, CA: Jossey-Bass.
Barrie, S. (2004) 'A research-based approach to generic graduate attributes policy'. *Higher Education Research and Development*, 23(3), 261–75.

Biggs, J. and Tang, C. (2007) *Teaching for Quality Learning at University: What the Student Does.* Buckingham, UK: Society for Research into Higher Education and Open University Press.

Bowden, J. and Marton, F. (1998) *The University of Learning: Beyond Quality and Competence in Higher Education.* London: Kogan Page.

Crawley, E.F., Malmqvist, J., Östlund, S. and Brodeur, D.R. (2007) *Rethinking Engineering Education: The CDIO Approach.* New York: Springer.

Dahler-Larsen, P. (2005) *Den rituelle reflektion – om evaluering i organisationer.* Odense, Denmark: Syddansk Universitetsforlag.

Edström, K. (2008) 'Doing course evaluation as if learning matters most'. *Higher Education Research and Development,* 27(2), 95–106.

Edström, K., Törnevik, J., Engström, M. and Wiklund, Å. (2003) 'Student involvement in principled change: understanding the student experience'. In C. Rust (ed.) *Improving Student Learning: Theory, Research and Practice. Proceedings of the 2003 11th International Symposium,* pp. 158–70. Oxford, UK: OCSLD.

Gibbs, G. (1992) *Improving the Quality of Student Learning.* Bristol, UK: Technical and Educational Services.

Gibbs, G. (1999) 'Using assessment strategically to change the way students learn'. In S. Brown and A. Glasner (eds.), *Assessment Matters in Higher Education.* Buckingham, UK: Society for Research into Higher Education and Open University Press.

Gibbs, G. (2010) *Dimensions of Quality.* York, UK: The Higher Education Academy.

Handal, G. (1996) *Studentevaluering av undervisning: Håndbok for lærere og studenter i høyere utdanning.* Oslo, Norway: Cappelen Akademisk Forlag.

Harvey, L. (2003) 'Student feedback [1]'. *Quality in Higher Education,* 9(1), 3–20.

Holmegaard, H.T., Ulriksen, L. and Madsen, L.M. (2010) 'Why students choose (not) to study engineering'. *Proceedings of the Joint International IGIP–SEFI Annual Conference 2010,* Trnava, Slovakia, 19–22 September 2010.

Hounsell, D. (2003) 'The evaluation of teaching'. In H. Fry, S. Ketteridge and S. Marshall (eds.), *A Handbook for Teaching and Learning in Higher Education: Enhancing Academic Practice.* London: Kogan Page.

Kember, D., Leung, D.Y.P. and Kwan, K.P. (2002) 'Does the use of student feedback questionnaires improve the overall quality of teaching?'. *Assessment and Evaluation in Higher Education,* 27(5), 411–25.

Kogan, M. (1990) 'Fitting evaluation within the governance of education'. In M. Granheim, M. Kogan and U.P. Lundgren (eds.), *Evaluation as Policy Making: Introducing Evaluation into a National Decentralised Educational System.* London: Jessica Kingsley Publishing.

Marton, F. and Säljö, R. (1984) 'Approaches to learning'. In F. Marton, D. Hounsell and N. Entwistle (eds.), *The Experience of Learning.* Edinburgh, UK: Scottish Academic Press.

Mazur, E. (1997) *Peer Instruction: A User's Manual.* Upper Saddle River, NJ: Prentice Hall.

Perry, W.G. (1970) *Forms of Ethical and Intellectual Development in the College Years: A Scheme.* New York: Holt, Rinehart and Winston.

Power, M. (1999) *The Audit Society: Rituals of Verification*, Oxford, UK: Oxford University Press.

Prosser, M. and Trigwell, K. (1999) *Understanding Learning and Teaching: The Experience in Higher Education*. Buckingham, UK: Society for Research into Higher Education and Open University Press.

Ramsden, P. (1992). *Learning to Teach in Higher Education*. London, Routledge.

Schreiner, C. and Sjøberg, S. (2007) 'Science education and youth's identity construction – two incompatible projects?'. In D. Corrigan, J. Dillon and R. Gunstone (eds.), *The Re-emergence of Values in the Science Curriculum*. Rotterdam, The Netherlands: Sense Publisher.

Seymour, E. and Hewitt, N.M. (1997) *Talking About Leaving: Why Undergraduates Leave the Sciences*. Boulder, CO: Westview Press.

Ulriksen, L., Madsen, L.M. and Holmegaard, H.T. (2010) 'What do we know about explanations for drop out/opt out among young people from STM higher education programmes?'. *Studies in Science Education*, 46(2), 209–44.

Weick, K. (1976) 'Educational organisations as loosely coupled systems'. *Administrative Science Quarterly*, 21(1), 1–19.

2

Advances in engineering education in Chile using student feedback

Mario Letelier, Rosario Carrasco, Daniela Matamala, Claudia Oliva, Doris Rodés and María José Sandoval

Abstract: Two main social forces are causing universities in Chile increase their effectiveness in translating market and political signals into actions in order to improve the quality and impact of the engineering education they provide. Since 2000 these forces have been expressed in government policies of programme accreditation and funding. This national context is exemplified through the experience at the University of Santiago of Chile. The expectations of the workplace with regard to the strengths and weaknesses of engineering education have been usually expressed through the desired learning outcomes of the programmes, many of which address attitudes rather than scientific or technological aspects. From this perspective, the actual developments in student feedback, considered as a significant resource for quality improvement, are discussed. Finally, some conclusions about structural factors that, in both the industrial sector and the universities, strengthen the achievement of the necessary consistency between the country's needs and university responses are drawn.

Key words: Chile, students, alumni, feedback, engineering.

Introduction

The main objective of this chapter is to present and analyse the current situation in the provision of feedback from engineering students in Chile. This is a snapshot of an evolving situation. In Chile, as in many other countries, education has been assigned great political priority as a development factor. This naturally leads to investment and innovation.

This chapter outlines the national context regarding student feedback, presents the case of the University of Santiago of Chile (USACH) and offers some conclusions, including a summary of the advances, challenges and reflections related to the relevance and complexity of making student feedback an effective tool to enhance the quality of their education.

The context of tertiary education in Chile

Chile has a population of 17 million people and has three types of tertiary institutions: universities, professional institutes and centres for technical education. Universities, as non-profit corporations, can offer all levels of programmes, from short technical programmes to doctoral ones. The total enrolment in tertiary education is around 850,000 students, a number that is continuously increasing.[1]

Most universities have engineering colleges, known as 'faculties' in Chile. The engineering programmes are very diverse. The length of their courses ranges from four to six years with variable scientific content. Six-year programmes are mainly 'civil engineering' programmes, where the word 'civil' is taken as a generic term meaning 'non-military'. These programmes have the strongest scientific basis, and constitute the main referent for this chapter.

Quality in the Chilean tertiary sector

Two main social forces are making universities in Chile increase their effectiveness in translating market and political pressures into actions aimed at improving the quality and impact of the engineering education they provide. These forces are expressed in government policies of programme accreditation, which started in Chile in 2000. This was accompanied by the funding of special projects, aimed at helping universities to overcome their academic weaknesses. Demands from the professional and industrial sectors have been complementary factors that have contributed to shaping the desired innovations.

In 1990, a licensing process for new private institutions at tertiary level was introduced. This process leads to full institutional autonomy. In 1999, the Ministry of Education started a pilot project aimed at establishing programme and institutional accreditation for autonomous institutions. One relevant outcome of this pilot project was the introduction of the Higher Education Quality Assurance Law, which came into force at the end of 2006. This law establishes that the National Commission for Accreditation (Comisión Nacional de Acreditación (CNA)) must conduct, among other activities, undergraduate programme accreditation, graduate programme accreditation, institutional accreditation, and the implementation of an information system (Comisión Nacional de Acreditación de Pregrado (CNAP), 2007). The CNA is part of the National System of Higher Education Quality Assurance (Sistema Nacional de Aseguramiento de la Calidad de la Educación Superior). It is backed by the National System of Higher Education Information (Sistema Nacional de Información de Educación Superior), which has integrated information related to enrolment, alumni, faculty, retention, finance, programme length, etc.

The present authors have been involved, at different levels, in both government-directed improvements and in exploring the market demands. This chapter describes some general patterns of student feedback management in Chilean universities that have emerged, mainly due to external pressures.

Developments at the national level are complemented by the institutional experience at the USACH. The USACH Centre for Research in Creativity and Higher Education has been in charge of facilitating programme accreditation and curricula development for several years, working closely with the local authorities in charge of this task.

External inputs coming from quality assurance activities and from the professional field have attempted to reflect the expectations of the workplace. There is wide agreement about the strengths and weaknesses of engineering education. These are expressed through learning outcomes, many of which address attitudes rather than scientific or technological aspects (Instituto de Ingenieros, 2005a; Brunner, 2008).

Student and alumni feedback in Chilean higher education

The Chilean higher education system, in particular the university sector, has suffered from the impact of the various forces demanding the use of student feedback in order to improve the teaching and learning in higher education. The greater visibility of students as actors providing feedback on the teaching and learning process has come in response to the exponential expansion and diversification of the higher education system, which has constantly increased in terms of enrolment, creation of private universities and campuses, as well as a proliferation of undergraduate and graduate programmes (OECD and World Bank, 2009).

Higher education policy agencies have read this process of massification as a signal for the need for quality improvement, especially among those institutions with very dubious academic performance (Letelier et al., 2009; Zapata and Tejeda, 2009).

Since the 1990s the institutional evaluation, and later the programme evaluation, included self-assessment processes that required gathering and analysing feedback from students, alumni and employers, which has been based on questionnaires. This is supposed to illustrate the programmes' academic strengths and weaknesses. Due to the expansion of accreditation experiences, these institutions have incorporated the revision of study programmes and the discussion of student outcome profiles in their agendas (Scharager and Aravena, 2010; Aedo, 2010).

Complementary to these accreditation procedures, another governmental initiative that has demanded more reliable data on student feedback comes from competitive funded projects, mainly MECESUP[2] funds. These have aimed to improve the quality of undergraduate and graduate programmes, research, innovation, and university management (División de Educación Superior, 2009; Canales et al., 2008).

Although there are no formal studies in the public domain concerning the variety of student feedback implemented throughout these projects, they are meant to provide an opportunity to gather feedback from students who should benefit from each project's innovations. Feedback is mainly collected through surveys that assess the students' overall satisfaction with the projects' outcomes and the application of tests of academic achievement.

At an institutional level, the most widely utilised form of student feedback has been student course evaluations. Although some universities had been conducting these before the introduction of government-directed self-assessment and accreditation obligations, others have only recently started to design and implement these government-directed quality assurance interventions. Despite Chilean universities being acquainted with feedback collection instruments, they face similar challenges to those mentioned in other higher education systems, such as (Gravestock and Gregor-Greenleaf, 2008):

- improving information and education of evaluation users of tested results;
- developing and testing effective means of reporting results and tools for interpretation (in relation to user needs);
- ensuring faculty and students are committed to the evaluation process;
- regular review of evaluation instruments based on institutional needs and goals and in relation to current research findings.

More recently, Chilean scholarship on student and alumni feedback has also increased due to research conducted by administrators and faculty members, which has focused on:

- the teacher–student relationship (Gallardo and Reyes, 2010);
- pedagogical knowledge and performance of future teachers (Echeverría, 2010);
- educational differences among recently graduated primary education teachers (Ruffinelli and Guerrero, 2009);
- motivation towards engineering and technological careers (Blázquez et al., 2009);
- perception of alumni, employers and other actors regarding higher education (Centro de Medicion MIDE-UC, 2008);
- educational and workforce pathways of doctoral graduates in science and engineering (Universidad Diego Portales y Universidad de Chile, 2010).

Student and alumni feedback in engineering

Given that Chilean universities tend to function independently of each other, and that there are confidentiality issues that protect the identity of individual students, it cannot be said that the following analysis depicts a comprehensive 'diagnosis' of student feedback practices among engineering programmes. Instead, it describes a limited number of these student feedback experiences with the intention of highlighting the main trends and considerations for future evaluations.

According to the experiences of five leading Chilean universities consulted during the research for this chapter, it can be stated that feedback processes have been implemented in recent years, although they have been collected for different purposes, such as self-assessment activities, MECESUP projects, and the Latin American Tuning Project[3] surveys.

In general, self-assessment procedures, as well as MECESUP projects, have provided a strong impulse to the exploration and implementation of student and alumni feedback, since these two initiatives are closely related (and are intended to be so). The MECESUP projects have helped to improve the critical dimensions of engineering programmes, and the self-assessment processes in institutions.

Student feedback has been developed in order to make decisions relating to the improvement in the quality of study programmes. Student feedback has highlighted, for example, that student drop-out rates are one of the most serious problems, followed by the extended length of study, or rigid curricular structures.

The most commonly used techniques, in both MECESUP and self-assessment processes, have been questionnaires that explore students' satisfaction with diverse educational dimensions, such as: the teaching and learning process, infrastructure and educational resources. As for alumni feedback, the questionnaires and interviews aim to examine their employment pathways, average time to obtain employment, average income, and the relationship between the work carried out and the field of study. Both face-to-face and electronic surveys have been utilised.

In general, the information gathered has been used to inform decision-making processes related to the revision of student outcome profiles (according to professional requirements) and the revision of study programmes, according to national and international accreditation criteria. As for the many impacts observed, the faculty consulted perceived positive effects, such as an increase in student retention rates in first-year courses, implementation of active learning methodologies, and improvement in the employability of graduate students.

Despite agreement among faculty that the information provided in student and alumni feedback is clear and can easily be utilised to improve the educational processes, faculty and university leadership continue to be reluctant to make more complex innovations based on this information.

As for the investment in student and alumni feedback, there is no policy on minimum spending on feedback activities. In the cases referred to in this chapter, it varies from less than 5 per cent up to approximately 25 per cent of the project's total institutional budget.

Neither MECESUP projects nor accreditation procedures have resulted in the development of systematic feedback procedures for engineering students and alumni. Student course evaluations continue to be practically the only systematic means of student feedback which face serious challenges related to, for example, a lack of published results, and hardly any use of the data to evaluate the process and instruments.

As for more sporadic experiences, the Latin American Tuning Project conducted a massive online survey in 18 countries during 2006 in order to explore the degree of importance and achievement of specific competencies in civil engineering. Faculty, students, employers and alumni were surveyed in those countries. More than 3,000 valid survey responses were analysed, and approximately 800 of those were collected in Chile. This allowed a comparative analysis of the most and least important specific competencies, according to the perceptions of the four groups of participants. The results have been discussed in various focus groups and elsewhere, and a book containing the main results was also published. Likewise, the MECESUP and self-assessment initiatives' data proved to be useful for the revision of study programmes, and provided diagnostic information for the elaboration of MECESUP projects.

Initiatives by the Institute of Engineers of Chile (IIC)

Further initiatives have been undertaken by the IIC, and one of its associated corporations, the Chilean Society for Engineering Education (SOCHEDI). The IIC is a 122-year-old non-profit corporation that aims to develop engineering and engineers in Chile. The IIC has commissioned several studies on the process of professional initiation of engineers in the work context. In Chile, universities provide academic degrees and professional titles. However, there is no professional licensing so far, although this may change in the near future. These studies were instigated as a result of realising that universities were not paying proper attention to the changing demands of the industry on graduating engineers.

Three studies conducted by IIC (Instituto de Ingenieros de Chile, 2005a; 2005b; 2010) collected opinions from employers, recent graduates and final-year students about job expectations and success. In the 2005

IIC study, responses from 39 companies, 211 alumni, and 684 students were obtained. Seven major faculties of engineering collaborated on this study. Questionnaires, meetings and focus groups were the main methods for collecting information. The most relevant findings are listed below.

1. In general, the scientific and technical education provided by the universities is valued, together with associated academic skills such as logical thinking, abstraction and systemic thinking.

2. Many general, or soft, skills are not acquired at an appropriate standard before entering the workplace. These include:
   - mastering the native language;
   - professional proficiency in English;
   - ample knowledge of computer technology;
   - global thinking;
   - team work.

3. Some higher-order technical skills are not being developed by the engineering curricula, according to the Chile's projected needs. The most frequently stated needs are:
   - effective use of engineering knowledge and methods to solve professional problems;
   - innovation based on scientific knowledge.

4. Development of some higher-order soft skills requires more attention and effort, such as:
   - professional responsibility and ethics;
   - leadership;
   - willingness to travel and to work in other countries and cultures;
   - interpersonal relationships.

The findings by the IIC presented educational challenges which the faculties of engineering accepted as requiring more attention. Complemented by other external inputs and by institutional educational projects, universities are modifying their curricula, fostering innovation and research on education, and creating support structures aided by the funds provided by the MECESUP.

It is the view of the present authors that employer and alumni feedback is currently given priority as a source of quality assurance in Chilean higher education. So far, student feedback does not appear to play an equally significant role. Most university managements tend to take the view that students are 'not mature enough' to provide reliable feedback related to the main educational objectives of the curricula.

On the other hand, the government creates policies expecting universities to increase their enrolment so that the growing demand for higher education is met. That demand comes mainly from sections of the population that formerly had little access to education due to a lack of financial means and poor academic background. As conditions improve, many universities face the problem of obtaining a better understanding of the learning capacities of a diversity of students. This is understood as a multi-faceted challenge that requires research related to admission regulations, learning processes and conditions, drop-out rates, motivation, intercultural factors and other aspects. In turn, student feedback on these aspects is required.

The case of USACH

USACH is a state-owned corporation created in 1947, by the fusion of several technical institutions that were distributed around the country. The oldest of these, the Escuela de Artes y Oficios, was founded in 1849. The total student enrolment at USACH stands at 16,500. Currently, 63 undergraduate programmes are offered, together with 14 doctoral, and 37 Master's programmes.

The USACH Faculty of Engineering is the largest in Chile, with an enrolment of 6,000 students. It has 21 undergraduate programmes, out of which nine are in civil engineering. These programmes are distributed among the following departments: electrical, civil projects, chemical, computing, geographical, industrial, mechanical, metallurgical and mining engineering.

USACH started a voluntary programme of self-assessment several years before the CNAP started the official accreditation of programmes. In part, this paved the way for participation in the official procedures. Notwithstanding, it would be an exaggeration to claim that there has not been considerable internal resistance. Until the end of the 1990s, USACH utilised a traditional approach to teaching, in which lecturers were the 'source of knowledge' and students the 'receptors', without seeking feedback from students.

During the first decade of the 2000s, USACH has achieved an internal consensus at all levels about implementing a quality cycle marked by clearly stated purposes, coherent activities and investments, measured results and impacts, feedback from key stakeholders, and actions to achieve improvements. A policy document was issued two years ago (University of Santiago of Chile, 2009) with the mission of fostering a quality culture and helping to monitor the application of the quality cycle to the already-defined main institutional processes. The present authors and others within the institution believe that this document is not sufficient to bring about a change in the internal culture of USACH. It is only a step in the direction of change and needs to be complemented by many other approaches and tools.

At most Chilean universities (including USACH), government requirements on accreditation have instigated the process of collecting feedback from students, alumni and employers and taking this feedback seriously. The involvement of USACH in national accreditation procedures started in 2002. Engineering programmes have also gradually entered the process. To date, 21 programmes have entered the process, and 13 of them have already been accredited.

Students are asked to evaluate the relevance of the curricula, faculty competence, resources and management, and other aspects. So far, about 5,000 students and 1,100 alumni have provided feedback. This experience has taught faculty departments several lessons. It was found that there was a general lack of information about alumni and it was difficult to reach them. Students were not necessarily familiar with concepts such as learning outcomes, professional relevance, etc., which would enable them to provide some in-depth input.

An important source of official information comes from the accreditation statements issued by the CNA. The statements for programmes include some statements about the conditions regarding the programme's operation, strengths and weaknesses, the number of accreditation years (these are in the range 2–7) and the improvements the programme must make.

Impact of feedback on study programmes

The most common way in which students participate, as providers of academic feedback, is as information providers, rather than participating in the full evaluation process. Faculty members are not prone to assign great value to student inputs. Typically, they distrust the students' capacity to evaluate the teaching. This is reflected in the following remark made in a CNAP evaluation of a civil engineering programme:

> ... even when students answer questionnaires, its results are not used as real feedback for improving the teaching–learning process. (Comisión Nacional de Acreditación de Pregrado, 2005)

On the other hand, students express their interest in the self-assessment process, but do not always show up when they have the opportunity to do so.

Despite this, student and graduate feedback has made an impact on some study programmes where positive changes in the curriculum were introduced on the basis of suggestions from students and alumni. For instance, in the case of civil electrical engineering, personal development and entrepreneurial courses have been incorporated in order to strengthen the development of soft skills for future engineers. Obligatory internships have also been included through formal agreements with private corporations, and the number of language courses in English has been increased in order to assure an adequate development of language skills in accordance with market expectations. It is important to note that another complementary force that made these innovations possible came from the recommendations of a MECESUP project implemented in the USACH Faculty of Engineering at the beginning of the decade, which achieved an important revision of the programme and graduate profiles. Thus, it is necessary to bear in mind the action of multiple initiatives that act synergistically rather than through isolated and sporadic actions.

On the other hand, the professional pathways of non-traditional engineering programmes have encountered problems regarding their legitimacy in the workforce context and have had to handle their students' uncertainty about their incorporation into the workforce. The programme of physical engineering had to handle these problems by organising, for example, meetings and events between graduate professionals and students, through which students gained a clearer perspective about the labour market and opportunities for graduates of this discipline, and possible internships for students in private organisations. The maturity of the engagement with student and alumni achieved in some of the engineering programmes has also been praised in their accreditation statements.

The USACH Faculty of Engineering received MECESUP funding for a project called 'Strengthening of structural capacities at the Faculty of Engineering and its applications to curricula innovation and didactics'. One of the objectives of this project was to create a new educational development unit. Its main goals were set as: curricular development, learning assessment, and alumni follow-up. This has made seeking student and alumni feedback and related activities more official.

Findings and discussion

Collecting student feedback in engineering in Chile is an ongoing process that has not yet achieved a satisfactory degree of maturity. Some conclusions and reflections based on the present authors' experience, internal and external reports and other literature are given below. They attempt to capture the current situation in Chile but also make some predictions.

Relevance of official pressures, policies and incentives

The change of focus from the teacher to the student is happening thanks to the combined pressures and orientations of many agents. The Ministry of Education fosters quality through the CNA and provides incentives by means of MECESUP funds. The professional field provides inputs related to the desired outcomes of university education. It seems safe to state that without these external forces, the traditional approaches to higher education would prevail. National policies in education are currently being revised, and greater pressure on universities regarding quality is expected.

Professional licensing

Among anticipated new policies in higher education and engineering is the enforcement of professional licensing. This means that university degrees may not directly entitle graduates to perform some engineering activities in the future. Licensing would promote closer links between the faculties of engineering and the professional field, alumni and the assessment agencies.

Value assigned to feedback

The faculties are moving from a minimal interest in student feedback to an increasing awareness of its relevance. Most large universities, USACH among them, have established institutional systems to survey students, alumni and employers in order to meet accreditation requirements. Their effectiveness has not yet reached the desired level. The difficulties associated with obtaining effective student feedback are becoming clearer as institutions gain experience. Such difficulties are outlined below.

Lack of faculty educational competencies

Despite many years of faculty training directed to developing competencies in curricular design, new teaching methods and learning assessment, it was found that lecturers lacked current pedagogical competencies. They were interested in discipline-specific areas such as transport phenomena, project management, mining processes or fluid mechanics; however, they had very little interest in the educational issues. Therefore their learning in that area was found to be relatively minimal, and changes in that area were not anticipated within a short time span.

Contrary to the government and university administrators' beliefs, the present authors are of the view that to convert a discipline-oriented academic into an educational expert is not a simple matter. For example, it is common to find that after many and expensive training periods, lecturers show a constant confusion of concepts such as abilities, skills, learning outcomes, relevance and quality. Therefore, we suggest that it is necessary to utilise other strategies in order to achieve the necessary innovations.

Communication challenges

In general, there is a significant generation gap between faculty and students in Chile. This brings with it a communication/language problem that has to be addressed. For example, students are often unfamiliar with the meaning of many concepts used when discussing educational subjects. Lecturers, therefore, need to develop a better understanding of the students' ways of thinking.

The student body is becoming increasingly diverse. Socio-economic background, motivations, gender, geographical barriers and ethnicity make it more complex for lecturers, used to more 'traditional'/uniform student cohorts, to build a common language with students.

Need for structural capacities

Some of the preceding remarks show the convenience of creating or reinforcing institutional capacities that do not depend upon the good will, motivations or time availability of faculty members. In Chile many universities, supported by MECESUP, are creating educational development units within faculties such as engineering, health and economics. These units, as in the case of USACH, aim to provide ongoing support to the subject-specific programmes in matters related to educational innovation.

Need for engineering education research

Many issues associated with quality assurance and student feedback require long-term institutional research studies (Research in Engineering Education Symposium, 2009). These would collect reliable institutional data that can be utilised in decision making on issues such as: student selection and retention, learning evaluation, curricular improvement, faculty training and selection, diversity, and discipline-specific long-term requirements.

Conclusions

As students become the centre of the educational process, many new challenges appear, some of which may require new institutional strategies. Students' ways of thinking, values, habits, self-concepts, needs (physical, psychological and social), diversity and academic background have to be seriously explored. Effective student feedback must consider these factors in order to duly capture students' responses.

Engineering faculties and even universities may not have all the competencies required to address these issues. The traditional academic culture places the professors at the centre of the educational stage, and they usually have a partial and old-fashioned image of learning processes. In Chile, professors belonging to disciplines such as natural sciences, engineering sciences, technology and business, typically look down on educational issues, such as those that are presently of prime relevance.

The national policies mentioned in this chapter are still being designed and implemented at a macro-level, with aligning funding to the country's educational needs as the first priority. As has been argued here, these policies are generating positive changes in the right direction.

A 10-year evaluation cycle has been completed, some results of which have been the subject of the present chapter. The authors have attempted to summarise the relevant impacts and their relationships with student feedback. It should be expected that the next cycle will address many of the issues mentioned in the first paragraph of the conclusions section of this chapter. Promising signals seem to be emerging from the government's statements about university funding. Most prominent is the retention issue which, in engineering (and undoubtedly other disciplines), necessarily leads to focusing attention on the learning process and the many factors associated with it.

Notes

1. The ratio of the number of people enrolled in tertiary education to the total population of Chile is 5 per cent.
2. MECESUP stands for Programa de Mejoramiento de la Calidad y la Equidad de la Educación Superior [Transl: Programme for Improving Higher Education Quality and Equity], which is funded by the Chilean Government.
3. The ALFA Tuning Latin America Project is an independent project, promoted and coordinated by universities in many different countries, both Latin American and European.

References

Aedo, A. (2010) 'Acreditación y Aseguramiento de la Calidad de los Egresados', Seminarios Internacionales CNED [Transl: Accreditation and quality assurance of alumni, International CNED seminars], vol. 13, pp. 145–51. Santiago, Chile: Consejo Nacional de Educación.

Blázquez, A., Álvarez, P., Bronfman, N. and Espinosa, J.F. (2009) 'Factores que influencian la motivación de escolares por las áreas tecnológicas e ingeniería' [Transl: Factors that influence students' motivation for technology and engineering areas]. *Revista Calidad de la Educación*, No. 31, pp. 45–64. Santiago, Chile: Consejo Nacional de Educación.

Brunner, J.J. (2008) 'Educación Superior y Mundo del Trabajo: Horizontes de Indagación' [Transl: Higher education and the workplace: horizons of inquiry]. *Revista Calidad en Educación*, No. 29, pp. 229–40. Santiago, Chile: Consejo Nacional de Educación.

Canales, A., De los Ríos, D. and Letelier, M. (2008) Proyecto Kawax-Bicentenario 'Comprendiendo la implementación de innovaciones educativas derivadas de programas MECESUP y CNAP para ciencia y tecnología' [Transl: Understanding the implementation of educational innovations derived from CNAP and MECESUP programmes for science and technology] Santiago, Chile: Editorial Universidad de Santiago de Chile.

Centro de Medición MIDE-UC (2008) Percepción de la calidad actual de los titulados y graduados de la educación superior chilena [Transl: Current quality perception of Chilean graduates]. Santiago, Chile: División de Educación Superior.

Comisión Nacional de Acreditación de Pregrado (2005) *Acuerdo de Acreditación N° 218* [Transl: Accreditation agreement No. 218]. Santiago, Chile: CNAP, Ministerio de Educación.

Comisión Nacional de Acreditación de Pregrado (2007) CNAP 1999–2007. El modelo chileno de acreditación de la educación superior [Transl: Chilean model of accreditation of higher education]. Santiago, Chile: CNAP, Ministerio de Educación.

División de Educación Superior (2009) Financiamiento por resultados MECESUP2 (préstamo 7317-ch) Avances de implementación [Transl: Funding according to MECESUP's results. Advances in implementation]. Ministerio de Educación, Programa MECESUP2, Unidad de Análisis y Convenios de Desempeño. Santiago, Chile.

Echeverría, P. (2010) 'El papel de la docencia universitaria en la formación inicial de profesores' [Transl: The university teaching role in initial teacher training]. *Revista Calidad de la Educación*, No. 32, pp. 149–65. Santiago, Chile: Consejo Nacional de Educación.

Gallardo, G. and Reyes, P. (2010) 'Relación professor–alumno en la universidad: arista fundamental para el aprendizaje' [Transl: Teacher–student relationship in university: the key part of the learning process]. *Revista Calidad de la Educación*, No. 32, pp. 77–108. Santiago, Chile: Consejo Nacional de Educación.

Gravestock, P. and Gregor-Greenleaf, E. (2008) 'Student course evaluations: research, models and trends'. Toronto, Canada: Higher Education Quality Council of Ontario.

Instituto de Ingenieros de Chile (2005a) Estudios sobre la Inserción Laboral de los Ingeniero Civiles en Chile [Transl: Studies on employability of civil engineers in Chile]. Santiago, Chile: Comisión de Educación.

Instituto de Ingenieros de Chile (2005b) Ética y Educación en Ingeniería [Transl: Ethics and education in engineering]. Santiago, Chile: Comisión de Educación.

Instituto de Ingenieros de Chile (2010) Análisis de la brecha existente entre la formación y las demandas del mundo laboral [Transl: Analysis of the gap between training and the demands of workplace]. Santiago, Chile: Comisión ingenieros jóvenes, nuevas perspectiva y proyectos laborales.

Letelier, M., Poblete, P., Carrasco, R. and Vargas, X. (2009) 'Quality assurance in higher education in Chile'. In A.S.P Patil and P.J.J Gray (eds) *Engineering Education Quality Assurance: A Global Perspective*, pp. 121–32. New York: Springer.

OECD and World Bank (2009) *La educación superior en Chile* [Transl: Higher education in Chile]. Paris: OECD.

Research in Engineering Education Symposium (2009) Held at Palm Cove, Queensland, Australia, 20–23 July 2009. Available online at: *http://rees2009. pbworks.com/w/page/5952012/Program%20with%20Papers* (accessed October 2011).

Ruffinelli, A. and Guerrero, A. (2009) 'Círculo de segmentación del sistema educativo chileno: destino laboral de egresados de Pedagogía en Educación Básica' [Transl: Segmentation of the Chilean education system: career prospects of graduates specialised in primary education]. *Revista Calidad de la Educación*, No. 31, pp. 19–44. Santiago, Chile: Consejo Nacional de Educación.

Scharager, J. and Aravena, M.T. (2010) 'Impacto de las políticas de aseguramiento de la calidad en programas de educación superior: un estudio exploratorio' [Transl: The impact of quality assurance policies in higher education programmes: an exploratory study]. *Revista Calidad de la Educación*, No. 32, pp. 15–42. Santiago, Chile: Consejo Nacional de Educación.

Universidad Diego Portales y Universidad de Chile (Centro de políticas comparadas de educación de la Universidad Diego Portales y Departamento de Ingeniería Industrial de la Facultad de Ciencias Físicas y Matemáticas de la Universidad de Chile) (2010) *Doctores en ciencia e ingeniería: trayectorias de estudio y situación laboral* [Transl: Science and doctors of engineering: study's pathway and career prospects]. Santiago, Chile: División de Educación Superior, Ministerio de Educación.

University of Santiago of Chile (2009) Direction of quality and systems: institutional model of quality assurance. Santiago, Chile: Pro-Rectory, University of Santiago of Chile.

Zapata, G. and Tejeda, I. (2009) 'Impactos del aseguramiento de la calidad y acreditación de la educación superior. Consideraciones y proposiciones' [Transl: The impacts of quality assurance and accreditation in higher education]. *Revista Calidad de la Educación*, No. 31, pp. 191–209. Santiago, Chile: Consejo Nacional de Educación.

3

Formative student feedback: enhancing the quality of learning and teaching

Kin Wai Michael Siu

Abstract: Formative student feedback is an issue on which researchers have increasingly focused, as both the learning and teaching of a course can be improved during the learning process. However, this type of continuous feedback is still paid less attention and is implemented less widely than are other learning and teaching enhancement methods. Based on a case study of an industrial and product design course in Hong Kong, this chapter first reviews the advantages of summative student feedback and identifies the common problems and limitations it presents. The chapter goes on to underline the importance of implementing a formative student feedback process. While identifying the difficulties inherent in this type of feedback, the chapter also discusses its possibilities, including recent educational changes and social expectations of the role of education.

Key words: Formative evaluation, student feedback, action research, quality of learning and teaching.

Introduction

Formative assessment of student performance is a technique that has been utilised since the 1960s. Its key value lies in its 'continuous' nature, whereby constructive reviews and recommendations can be provided to students throughout the learning process. However, this type of assessment has been utilised to a far lesser extent than summative student feedback to enhance the quality of learning and teaching. Most of the time, students provide feedback on learning content and teaching quality only at or after the end of the learning process. This type of summative evaluation of overall learning and teaching quality is limited to a review of the course itself, before recommendations on future iterations of the same and similar courses are made. When structured in this way, the learning and teaching of the course cannot be improved during the learning process. Taking Hong Kong as a case study, this chapter explores how formative student feedback can enhance the quality of learning and teaching in a constructive and formative way in the course of the teaching and learning process. The chapter first reviews the common problems and limitations associated with collecting and utilising feedback from engineering students in Hong Kong. Student feedback is generally collected only when the learning process has been completed. This one-off evaluation serves as a final evaluation rather than as formative feedback that can inform teaching practice as the course is delivered. Further, many students are reluctant to provide, or are not interested in, providing feedback after their unit or programme has finished. Based on a case study of an industrial and product design course forming part of an engineering programme, this chapter identifies and discusses the advantages, difficulties and possibilities presented by the implementation of formative student feedback practice. The discussion is situated in the learning and teaching culture particular to Hong Kong.

Case study: engineering education at the Hong Kong Polytechnic University

A continuous study on matters related to student feedback was conducted in Hong Kong in the period 2007 to 2010. The major objective of the study was to review the effectiveness of student feedback and its impact on improving learning and teaching. A case study (i.e., qualitative) approach was adopted to gain an in-depth understanding of the topic. An industrial and product design course offered by the School of Design with another engineering department at the Hong Kong Polytechnic University was selected as the course for the study. The study was conducted by a researcher from the school, with the participation of full-time, part-time and visiting lecturers and other tutors involved in the teaching and coordination of the course over the course of the study. Students were invited to participate in the study on a voluntary basis.

The study involved the following research activities:

- review of student feedback objectives and the forms in which such feedback was collected;
- review of student feedback collected;
- observation of learning and teaching activities in class;
- interviews with selected students;
- interviews with teachers (full-time, part-time and visiting lecturers and/or other tutors);
- interviews with programme and course coordinators and other related administrative staff.

To provide constructive insights into how to improve the learning and teaching of the course, the study utilised the 'action research' method. 'Action research' is a term first coined by Kurt Lewin in about 1944. He then outlined the method in a paper (Lewin, 1946). He described the method as a spiral of steps, each of which is composed of a circle of planning, action and fact-finding about the result of the action (Siu, 2000, 2007; Somekh, 2006).

Stage I of the present study was conducted in the 2007–2008 academic year. Student feedback on the course was collected in line with common practice for other courses throughout the university. That is, student feedback was collected only at the end of a course (including via the final assessment). This feedback was collected through a questionnaire focusing on the following aspects:

- rating of the course arrangements (e.g. learning and teaching environment, support from the general office);
- rating of the course contents (including assignments and assessments);
- rating of teacher performance (including tutors);
- other comments related to the course and the teacher(s) (including tutors).

Based on the findings made in Stage I, the form in which student feedback was collected was improved in the 2008–2009 academic year (Stage II) and the same process was repeated in Stage III (the 2009–2010 academic year). The findings and adjustments – actions, spiral of steps – made during the three stages are presented and discussed in the following section.

Summative student feedback

Advantages

The Stage I findings showed that the student feedback process conducted only at the end of the course was mainly summative in nature (for further discussion of summative evaluation and assessment, see Gioka, 2008; Glickman et al., 2009; Morrow, 2005; Patton, 2011; West, 1975). In general, summative student feedback has several key objectives and advantages.

First, collecting one-off, summative student feedback at the end of a course is easy and convenient in terms of planning and implementation, particularly when gauged by time required and human and other resources employed (McDowell, 2008; Morrow, 2005; Siu, 1998, 2000). It is easy and consistent to carry out collection of such a feedback for different programmes and courses. This is also why this form of student feedback has been widely adopted by many universities. Furthermore, consistency is now an important requirement and concern in course evaluation. People responsible for evaluating courses generally prefer to eliminate all non-standard factors and items, including non-standard tools for collecting feedback and evaluation, collected data, and methods of analysis which are relatively more difficult to manage (McDowell, 2008; Rudney and Guillaume, 2003; Siu, 2007).

One advantage of adopting a standardised summative student feedback format is that it enables comparison between different programmes, courses and teachers (Davidson-Shivers, 2006; Kimball, 2001; King, 2003; McDowell, 2008; Rudney and Guillaume, 2003; Siu, 2000; West, 1975). Although a number of researchers have argued that it is inadequate to evaluate and compare different programmes and courses using the same means of evaluation (such as standardised, inflexible methods and tools), many programme planners, coordinators and lecturers still expect to use more standardised means of assessment (O'Donoghue, 2010; McDowell, 2008; Rudney and Guillaume, 2003; Siu, 1998, 2000). To eliminate discrepancies due to differences in tools and a lack of quantitative elements, summative student feedback conducted at the end of a course is generally considered the most reasonable, rational and widely applicable approach (for standards of evaluation and assessment, see Aylett and Gregory, 1996; Cascallar and Cascallar, 2003; Ellis, 1993; McDowell, 2008; Thackwray, 1997).

Another advantage of obtaining summative student feedback at the end of a course relates to time and post-evaluation management issues. First, there is no pressure to take prompt or immediate action (i.e., data analysis, feedback and action for change and improvement) after feedback is collected (Siu, 2007). For example, this study highlighted the intensive nature of today's programme structures and demonstrated that administrative and academic staff bear a heavy workload. According to a programme coordinator interviewed for this study, collecting student feedback at the end of a course implies that it is not so urgent to do so, or at least that academic and administrative staff are not under pressure to take immediate action to analyse data and make suggestions for improvement. In many Asian universities, a department may now run a large number of concurrent programmes (covering more than a hundred courses), particularly due to the rapid increase in self-financed programmes. Further, as pointed out by programme coordinators and teachers interviewed in this study, analysis and follow-up action immediately after the student feedback has been collected is nearly impossible.

Common problems and limitations

Summative student feedback entails a number of problems and limitations that critically affect the quality of learning and teaching. In practice, some of its advantages and benefits can also give rise to problems and limitations. While summative student feedback can be considered a type of 'outcome evaluation' (Gioka, 2008), its key limitation is that this type of feedback cannot provide prompt and timely guidance on action leading to immediate improvements (Siu, 2007).

Although consistency in collecting and analysing student feedback is important, collecting student feedback in a rigid and consistent fashion may sometimes have its disadvantages. For example, it may result in student feedback collection becoming a routine administrative requirement, rather than a constructive form of action taken to improve the programme and course evaluation process and inform learning and teaching activities. This is contrary to the core objective of collecting student feedback. In addition, different programmes and courses require different learning and teaching environments, or different learning and teaching objectives that may be 'flattened' by applying a universal evaluation approach. It is neither effective nor constructive to utilise a standardised evaluation approach regardless of the nature of the programme or course (Davidson-Shivers, 2006; McDowell, 2008).

As was pointed out by a teacher who was interviewed, student feedback provides ratings in different areas related to programme planning, course arrangements and contents, learning and teaching activities, and teacher performance. Summative student feedback generates useful and constructive comments, opinions, and suggestions. It still fulfills the important purpose of guiding 'future' improvements (O'Donoghue, 2010; Siu, 2000). However, the major limitation of one-off summative feedback provided at the end of the course is that it cannot inform constructive action that might be taken to improve the teaching of the course before the course is finished. First, the function of student feedback is restricted, or it cannot be maximised. Moreover, as discussed above, collecting summative student feedback fails to provide course leaders and teachers with immediate feedback on how to make immediate improvements to the programme (Gioka, 2008). Further, although students who give feedback have an opportunity to raise their concerns, they do not benefit from their own advice. Finally, teachers cannot benefit from students' suggestions (i.e. make any improvements) before the course has been completed.

In recent years, more researchers and educators have emphasised that learning and teaching is a continuous process. In particular, learning objectives are now expected to centre more on the learning process, rather than solely on the final output (Coles, 1997; Dew and Nearing, 2004; Fisher, 2003). Thus, it is important to obtain feedback on different stages of a course; this goal cannot be attained through summative student feedback (Gioka, 2008).

A key finding of this study is that the motivation for students to give feedback was low. Several of the students who were interviewed indicated that they did not take the feedback questionnaires seriously. This was also reflected in the student feedback questionnaires, where students rarely provided written comments in the 'other comments' section of the questionnaire. One common perception among students was that this type of summative evaluation did not provide any practical benefit for their learning of the course they had just completed. Some of the students also commented that they did not believe the programme and course coordinators and teachers would take their feedback seriously. Thus, students placed a low value on feedback questionnaires.

In the same way, the low level of student motivation to give feedback also implies that the feedback provided via students' comments, opinions and suggestions was low in quality. A ripple effect then occurred, in that the programme and course coordinators, and especially the teachers, did not take student feedback seriously. Course teachers were most concerned about whether they had reached the rating thresholds applied by the school and university. Many of the teachers also admitted that they seldom read other written comments provided by students unless they saw that the rating of their teaching performance was low or sensed that the student had raised a complaint.

In summary, it would be unfair to say that summative student feedback serves no useful purpose. One teacher interviewed for this study indicated that he paid attention to negative comments made by students to improve his future teaching. Nevertheless, summative feedback cannot bring about any change to the course just completed or directly benefit those students who supply comments.

Formative feedback

Formative student feedback involves a continuous evaluation process in which action is pivotal to its success (Cox, 1974; Donaldson and Scriven, 2003; George and Cowan, 1999; Nilson, 2010; Siu, 2007; Stockley, 2006). In Stages II and III of this study, student feedback was collected not only at the end of the course, but also as it was delivered.

In Stage II, an additional student feedback questionnaire was distributed in the mid-term of the course (i.e. in the 7th week of a 14-week term). The format and contents of the questionnaire were similar to those of the questionnaire distributed at the end of the course. The major reason for employing two similar student feedback questionnaires for the middle stage and the end of the course was to obtain comparative data on teacher performance and comments on course contents.

In Stage II, before the course commenced, the teachers involved in the course were provided with student feedback collected in Stage I. After the mid-stage student feedback exercise, the teachers involved in the course were requested to note and attempt to improve their teaching according to the feedback provided.

The core arrangements for Stage III (including the questionnaires) were the same as those adopted in Stage II. The questionnaire items were the same as those included in the questionnaires distributed in Stages I and II. Based on the student feedback and case study experience gained in Stage II, another student feedback exercise was carried out in Stage III. Several small student group meetings were organised to collect comments on the course. Instead of setting up particular time slots to collect student feedback, feedback was collected at the end of some tutorials. This was a more effective way of collecting student feedback, as students were more willing to attend small group meetings because they did not need to spend extra time providing their feedback.

Advantages

The major advantage of formative student feedback is that it provides an opportunity for improvements to be made during the learning and teaching process (Stockley, 2006; Tessmer, 1993). It can be considered a process where both learning and teaching activities and the evaluation and improvement process can be run in parallel (Thackwray, 1997). In Stages II and III, due to the feedback collected during the course (i.e. through the mid-stage questionnaire and small group comments), teachers had an opportunity to make improvements to course contents and their teaching performance. This was reflected in the final student feedback, where students commented that their experience of the course improved after they had provided mid-stage feedback.

Formative student feedback also has the advantage that learning and teaching activities can be improved according to students' changing needs (George and Cowan, 1999; Gioka, 2008; Holmes and Brown, 2000; Nilson, 2010; Tessmer, 1993). Because coordinators and teachers may not be familiar with their students' backgrounds and learning needs (e.g. age, study experience or expectations of the course) before the course begins, mid-stage feedback may assist them in addressing this gap in their knowledge (Holmes and Brown, 2000; Siu, 1998). Thus, the formative student feedback collected during the course in this study provided teachers with the flexibility and opportunity they needed to adjust their teaching content and methods. For example, in Stage II, the teacher originally assigned students a design project utilising a particular software application that had received highly positive ratings from students in previous academic years (i.e. reflected in Stage I feedback). However, in the mid-stage informal small group meetings, some of the Stage II students informed the teacher that they would prefer to learn how to use a new software application. The teacher therefore modified the project requirements and allowed the students to select their preferred software application in the design implementation phase. As was also pointed out by Andrade and Cizek (2010), Behrendt (2001) and Mann (2006), this type of motivation generated through formative evaluation is important for effective learning (Siu, 1998; Stockley, 2006; Van Evera, 2004). Flexibility and opportunity for change are also major objectives of formative student feedback.

Moreover, students indicated their appreciation of the opportunity to provide informal feedback in small group meetings, and most students were more willing to give their comments on the course. The major reason for this was that students realised that the teachers were taking their feedback seriously and wanted to improve the quality of the course, rather than treating the feedback process as a routine exercise required by the school and university. This became more obvious in the mid-stage student feedback and small group comments.

This increased student motivation was reflected in students providing more (in quantity) and more detailed (in quality) comments in the mid-stage feedback exercise, and teaching practice also improved as a result of the student feedback (compared to Stage I). The qualitative change in teaching practice was highlighted by several of the students who were interviewed. The students perceived that teachers prepared much more thoroughly for the mid-stage feedback exercise, and that they took student feedback more seriously. The arrangements teachers made were different both from those made in the past and from those implemented for other courses. Thus, instead of giving only a marginally positive rating for each item as was the general practice in the past, the students showed they were more serious in providing feedback on each individual item. This was particularly clear because in Hong Kong and many Asian countries, students seldom take this type of student feedback exercise seriously. In treating it as a routine course evaluation exercise, many students seldom give negative and critical comments about a teacher unless the teacher's performance or attitude is very bad.

As discussed above, this higher level of motivation to provide feedback can result in more and more detailed comments being obtained from students (Nilson, 2010). The questionnaire review conducted in this study showed that students were more willing to provide comments on how to improve the course and teaching performance. As indicated by the teachers, more serious comments gave clearer and detailed guidance to course coordinators and teachers on how to improve the course content and enhance learning and teaching activities.

The previous discussion demonstrates that the advantages and value of formative student feedback lie not only in the objective and content of feedback on how to enhance the course, but also in the creation of a culture of providing continuous feedback for improvement.

Difficulties

Collecting formative student feedback in Stages II and III was more complicated than collecting summative feedback in Stage I. The coordinators and teachers who were interviewed, noted that the process consumed time and resources and that the teachers had to take on additional responsibilities in terms of preparation, implementation, data analysis and follow-up.

In comparison with Stage I, teachers could not rely on a ready-made and standardised means of collecting student feedback: the nature of collecting formative student feedback required them to make adjustments to the ways in which feedback was collected. Despite their significant benefits, the formative student feedback exercises conducted in Stages II and III created significantly increased the teachers' workloads.[1] As noted by a teacher involved in Stage III teaching, the extra workload became even more obvious when improvements were required and student feedback was translated into action. The teacher pointed out that improvements were required not only to the course contents and learning and teaching activities, but it also required at least fine-tuning of the evaluation methods and contents for the next stage. The same teacher also indicated that this increased workload would have to be maintained if formative evaluation was to remain meaningful.

Data analysis is another major part of the formative student feedback process that presents difficulties for education professionals (Andrade and Cizek, 2010; Davies, 2003). Its challenge lies in the greater potential of formative student feedback to include qualitative components in the matters evaluated (Mann, 2006; Siu, 1998). This then implies a need for analysis of qualitative data, creating a potential increase in workload. Teachers interviewed for this study indicated that this type of data, encompassing diverse and sometimes contradictory comments, was not easy to analyse and distill into suggestions for improvement. Further, the data had to be analysed and suggestions promptly acted on. As indicated in recent research, this difficulty is also the reason why many programme and course coordinators and teachers refuse to utilise formative evaluation and assessment despite its proven advantages (Andrade and Cizek, 2010; Cascallar and Cascallar, 2003; Ellis, 1993; Rudney and Guillaume, 2003; Siu, 2007; Van Evera, 2004). This reluctance is also due to the fact that academic programmes in Asia are progressively intensifying the pressure on academics (Siu, 1998; Burden and Byrd, 2010; Holmes, 2009; Partin, 2009).

Possibilities

Although there are a range of difficulties involved in implementing formative student feedback, enhancing the quality of learning and teaching by applying a continuous and action-based evaluation programme cannot be overlooked as an option (Aylett and Gregory, 1996; Ellis, 1993; Donaldson and Scriven, 2003; Massy et al., 2007; Patton, 2011; Stockley, 2006). Particularly in recent years, educational changes and social expectations on the role of education have led more educators and members of the public to recognise the importance of formative student feedback and consider how it can be effectively implemented (Siu, 2007). This type of recognition and consideration is critically important as a driving force for the government, related education sectors, and the public to agree to put more resources and effort into promoting and utilising formative student feedback and to take action to achieve this end.

Education reforms differing in nature and scale have been implemented worldwide since the mid-1990s, particularly in the Asian region and in developing countries where policymakers have recognised the importance of good quality education. Among all the aspects of such reforms, quality assurance in programme planning, implementation, and evaluation has been highlighted as a key priority (Mok and Chan, 2002; Neave, 2000; Postiglione, 2002; Salmi, 2002). As discussed above, formative student feedback is valuable in enhancing the quality of learning and teaching activities. Thus, education reforms may become a force that pushes more academics and others to give serious consideration to the need for formative student feedback.

The recent emphasis on student-centred learning and outcome-based learning has also made it possible to promote and implement formative student feedback programmes (Nilson, 2010). This is because student feedback has become one of the key factors in education quality assurance. On the programme and course levels, it is important that learning and teaching are continuously evaluated. An action approach to collecting student feedback and making improvements is considered appropriate and effective (Stockley, 2006).

Changes in, and pressures on, higher education systems around the world have intensified through the internationalisation and globalisation of education programmes, with the greater number of students enrolling from different countries and regions making the situation even more complicated (Jiang, 2008). To follow these 'market' trends and needs, updated and accurate student feedback is crucial to the success of such programmes and courses. As discussed above, formative student feedback has the advantage of facilitating both evaluation and improvements in a prompt and effective manner. The continuous and action-based nature of formative student feedback can assist in adjusting to the changing and varied needs and preferences of diverse student cohorts. By utilising formative student feedback, problems and deficiencies in learning and teaching matters can be resolved at an early stage to prevent their accumulation (Massy et al., 2007; Tessmer, 1993).

A majority of the programme coordinators and teachers interviewed for this study believed that a greater proportion of academics were currently willing to take an active role in improving the quality of their teaching. In other words, more coordinators and teachers recognised the importance and practical value of obtaining student feedback, and thus of putting more effort into collecting and utilising such feedback.

In recent years, more Asian students have started raising their concerns, including suggestions for improvement. This has meant a change from the past, when many Asian students were reluctant to voice their views and comment on learning and teaching matters. Lin (2004) has also identified this as the reason Asian students found it difficult to adapt to foreign (particularly Western) learning cultures when they studied abroad. This change in learning styles among Asian students in recent years has enabled coordinators and teachers to introduce formative student feedback. Students' greater willingness for, and more active participation in, providing their feedback can also provide a greater motivation for academics to improve their teaching.

Another practical and positive shift in promoting and implementing formative student feedback is the increase in general support for collecting student feedback (Siu, 2007). Although many administrators and teachers have continuously complained about resource cuts or shortages in university education, the funding provided to support quality assurance has actually increased in recent years (Brown, 2004; Gokulsing, 2008; Higher Education Quality Council, 1995a). Taking Hong Kong as an example, not only has the amount spent on education programme quality assurance increased, but so too has funding for audits conducted by government councils, committees and international departments of universities. Student feedback was collected in a piecemeal manner in the past, whereas more universities in Hong Kong have now set up central quality assurance units. General and teaching support staff now assist with the student feedback collection process. All of these changes made in recent years have helped to eliminate some of the difficulties encountered in coordinating the student feedback collection process (Higher Education Quality Council, 1995b; Latchem and Jung, 2010; Williams, 2008).

Conclusions

Recent years have seen an increase in the use of formative student feedback for learning and teaching quality assurance purposes. Formative student feedback has been given a high priority. However, many programme coordinators and front-line teachers remain reluctant to utilise formative student feedback. Many of them still see student feedback as a routine quality assurance requirement designed to meet administrative and public obligations, instead of as a constructive means of providing continuous up-to-date practical benefits to both students and teachers.

Based on a case study undertaken in Hong Kong, this chapter identifies and discusses some of the key advantages, difficulties and possibilities presented by formative student feedback. In terms of its advantages, formative student feedback allows programme coordinators and teachers to make adjustments to learning and teaching activities. Students are given a greater opportunity to raise their views and concerns, and are more motivated to do so. This can lead to the development of more positive views among students on the merits of the evaluation process, and can further encourage students to give a greater number of more detailed comments and suggestions on how to improve learning and teaching activities. In this sense, the student feedback process ought to be considered a constructive source of, and driving force for, improvements in teaching and learning practice, particularly when carried out on a prompt and continuous basis.

The difficulties encountered in collecting formative student feedback are both directly and indirectly related to their advantages. Four aspects of student evaluations – preparation, implementation, data analysis and follow-up – are difficulties that generally discourage teachers from carrying out such evaluations in a continuous and active manner. Above all, time and manpower resources are crucial. Another major difficulty is variation in the degree of willingness among administrators and teachers to recognise the importance of formative student feedback and commit to conducting evaluations.

In recent years, the general public has put greater pressure on higher education institutions to demonstrate improvements in the quality of education. Students have taken a more active role in learning and teaching. As discussed above, the lack of interest in implementing formative student feedback programmes persists. Students' active participation and administrative support can motivate teachers to carry out formative evaluations. However, obtaining and reacting to student feedback requires prompt and continuous action.

Acknowledgements

The author would like to acknowledge the resources provided by the Hong Kong Polytechnic University and the Asian Scholarship Foundation to support the study discussed in this paper. He would also like to thank colleagues from the School of Design who participated in the study and provided useful information and constructive comments on the paper. He further expresses his thanks to the researchers and visiting scholars of the Public Design Lab for their assistance in the study.

Note

1. In the past, the student feedback collection process employed in the university was simple. General staff were assigned to deliver and collect questionnaires to and from students during the feedback exercise. All of the collected questionnaires would be passed to a central department in the university for standardised data analysis to be carried out, and the results of this analysis would ultimately be distributed to the relevant teachers and coordinators. The entire process was routine and standardised across the university.

References

Andrade, H.L. and Cizek, G.J. (eds.) (2010) *Handbook of Formative Assessment*. New York: Routledge.

Aylett, R. and Gregory, K. (eds.) (1996) *Evaluating Teacher Quality in Higher Education*. London: Falmer Press.

Behrendt, H. (ed.) (2001) *Research in Science Education: Past, Present, and Future*. Boston, MA: Kluwer Academic.

Brown, R. (2004) *Quality Assurance in Higher Education: The UK Experience Since 1992*. London: RoutledgeFalmer.

Burden, P.R. and Byrd, D.M. (2010) *Methods for Effective Teaching: Meeting the Needs of All Students* (5th edn.). Boston, MA: Allyn and Bacon.

Cascallar, A. and Cascallar, E. (2003) 'Setting standards in the assessment of complex performances: the optimized extended-response standard setting method'. In M. Segers, F. Dochy and E. Cascallar (eds.) *Optimising New Modes of Assessment: In Search of Qualities and Standards*, pp. 247–66. Boston, MA: Kluwer Academic.

Coles, M. (1997) 'Curriculum evaluation as review and development: the curriculum leader's role in creating a community of enquiry'. In M. Preedy, R. Glatter and R. Levačić (eds.) *Education Management: Strategy, Quality, and Resources*, pp. 113–26. Buckingham, UK: Open University Press.

Cox, R. (1974) *Formative Evaluation: Interpretation and Participation*. London: University Teaching Methods Unit, University of London Institute of Education.

Davidson-Shivers, G.V. (2006) *Web-based Learning: Design, Implementation, and Evaluation*. Upper Saddle River, NJ: Pearson/Merril Prentice Hall.

Davies, L.M. (2003) *Monitoring and evaluating adult education programs in the District of Columbia*. Unpublished thesis. Amberst, MA: University of Massachusetts Amherst.

Dew, J.R. and Nearing, M.M. (2004) *Continuous Quality Improvement in Higher Education*. Westport, CT: Praeger.

Donaldson, S.I. and Scriven, M. (2003) *Evaluating Social Programs and Problems: Visions for the New Millennium*. Mahwah, NJ: Lawrence Erlbaum.

Ellis, R. (1993) *Quality Assurance for University Teaching*. Buckingham, UK: Society for Research into Higher Education and Open University Press.

Fisher, M.M. (2003) *Designing Courses and Teaching on the Web: A 'How to' Guide to Proven, Innovative Strategies*. Lanham, MD: Scarecrow Press.

George, J.W. and Cowan, J. (1999) *A Handbook of Techniques for Formative Evaluation: Mapping the Student's Learning Experience*. London: Kogan Page.

Gioka, O. (2008) 'Teacher or assessor? Balancing the tensions between formative and summative assessment in science teaching'. In A. Havnes and L. McDowell (eds.) *Balancing Dilemmas in Assessment and Learning in Contemporary Education*, pp. 145–56. New York, NY: Routledge.

Glickman, C.D. Gordon, S.P. and Ross-Gordon, J.M. (2009) *Supervision and Instruction Leadership: A Developmental Approach*. Boston, MA: Allyn and Bacon.

Gokulsing, K.M. (2008) *The New Shape of University Education in England: Interdisciplinary Essays*. Lewiston, NY: Edwin Mellen Press.

Higher Education Quality Council (1995a) *Managing for Quality: Stories and Strategies: A Case Study Resource for Academic Leaders and Managers*. London: Higher Education Quality Council.

Higher Education Quality Council. (1995b) *A Quality Assurance Framework for Guidance and Learner Support in Higher Education: the Guidelines*. London: Higher Education Quality Council.

Holmes, A. and Brown, S. (2000) *Internal Audit in Higher Education*. London, Kogan Page.

Holmes, E. (2009) *The Newly Qualified Teacher's Handbook* (2nd edn.). London: Routledge.

Jiang, X.P. (2008). *Why Interculturalisation? A Response to the Internationalisation of Higher Education in the Global Knowledge Economy*. Rotterdam, The Netherlands: Sense Publishers.

Kimball, S.M. (2001) *Innovations in teacher evaluation: case studies of two school districts with teacher evaluation systems based on the framework for teaching*. Unpublished thesis. Madison, WI: University of Wisconsin-Madison.

King, K.P. (2003) *Keeping Pace with Technology: Educational Technology that Transforms*. Creskill, NJ: Hampton Press.

Latchem, C. and Jung, I. (2010) *Distance and Blended Learning in Asia*. New York: Routledge.

Lewin, K. (1946) 'Action research and minority problems'. *Journal of Social Sciences*, 2(4), 34–46.

Lin, C.Y. (2004) *Taiwanese students in a United States university: expectations, beliefs, values, and attitudes about learning and teaching*. Unpublished thesis. University Park, PA: Pennsylvania State University.

Mann, B.L. (2006) *Selected Styles in Web-based Educational Research*. Hersey, PA: Information Science Publishers.

Massy, W.F., Graham, S.W. and Short, P.M. (2007) *Academic Quality Work: A Handbook for Improvement*. Bolton, MA: Anker Publishers.

McDowell, L. (2008) 'Students' experiences of feedback on academic assignments in higher education: Implications for practice'. In A. Havnes and L. McDowell (eds.) *Balancing Dilemmas in Assessment and Learning in Contemporary Education*, pp. 237–50. New York, NY: Routledge.

Mok, K.H.J. and Chan, D.K.K. (2002) *Globalization and Education: the Quest for Quality Education in Hong Kong*. Hong Kong: Hong Kong University Press.

Morrow, J.R. (2005) *Measurement and Evaluation in Human Performance* (3rd edn.). Champaign, IL: Human Kinetics.

Neave, G. (2000) *The Universities' Responsibilities to Society: International Perspectives*. Amsterdam, The Netherlands: Pergamon.

Nilson, L.B. (2010) *Teaching at its Best: a Research-based Resource for College Instructors*. San Francisco, CA: Jossey-Bass.

O'Donoghue, J. (2010) *Technology-supported Environments for Personalized Learning: Methods and Case Studies*. Hersey, PA: Information Science Reference.

Partin, R.L. (2009) *The Classroom Teacher's Survival Guide: Practical Strategies, Management Techniques, and Reproducibles for New and Experienced Teachers* (3rd edn.). San Francisco, CA: Jossey-Bass.

Patton, M.Q. (2011) *Developmental Evaluation: Applying Complexity Concepts to Enhance Innovation and Use*. New York: Guilford Press.

Postiglione, G. (2002) 'Chinese higher education for the twenty-first century: Expansion, consolidation, and globalization'. In D.W. Chapman and A.E. Austin (eds.) *Higher Education in the Developing World: Changing Contexts and Institutional Responses*, pp. 149–66. Westport, CT: Greenwood Press.

Rudney, G.L. and Guillaume, A.M. (2003) *Maximum Mentoring: an Action Guide for Teacher Trainers and Cooperating Teachers*. Thousand Oaks, CA: Corwin Press.

Salmi, J. (2002). 'Higher education at a turning point'. In D.W. Chapman and A.E. Austin (eds.) *Higher Education in the Developing World: Changing Contexts and Institutional Responses*, pp. 23–44. Westport, CT: Greenwood Press.

Siu, K.W.M. (1998) 'The evaluation of higher education in a compressed spatial and temporal world'. *Education Today*, 48(1), 9–13.

Siu, K.W.M. (2000) 'Re-construction of learning space for design education'. *Design and Education*, 8(1), 20–8.

Siu, K.W.M. (2007) 'Balance in research and practice: critical reform of research studies in industrial and product design'. *Global Journal of Engineering Education*, 11(1), 15–27.

Somekh, B. (2006) *Action Research: a Methodology for Change and Development*. Maidenhead, UK: Open University Press.

Stockley, D. (2006) 'Strategies to collect and use online student feedback: Improving teaching through formative evaluation'. In B.L. Mann (ed.) *Selected Styles in Web-Based Educational Research*, pp. 246–59. Hersey, PA: Information Science Publishers.

Tessmer, M. (1993). *Planning and Conducting Formative Evaluations: Improving the Quality of Education and Training*. London: Kogan Page.

Thackwray, B. (1997) *Effective Evaluation of Training and Development in Higher Education*. London: Kogan Page.

Van Evera, W.C. (2004) *Achievement and motivation in the middle school science classroom: the effects of formative assessment feedback*. Unpublished thesis. Fairfax, VA: George Mason University.

West, R.W. (1975) *The Summative Evaluation of Curriculum Innovations*. Brighton, UK: University of Sussex Education Area.

Williams, B.C. (2008) *Preparing Effective Teachers of Reading: Putting Research Findings to Work for Student Learning*. New York: Peter Lang.

4

Role of soft skills in engineering education: students' perceptions and feedback

Mousumi Debnath, Mukeshwar Pandey,
Neelam Chaplot, Manohar Reddy Gottimukkula,
Pramod Kumar Tiwari and Suresh Narain Gupta

Abstract: It can be argued that soft skills are vital in virtually every workplace today, regardless of the domain. This chapter outlines research undertaken to analyse the role of soft skills in engineering education. The study was conducted at the Jaipur Engineering College and Research Centre, a leading technical institute in India. In this study, a personal report of communication apprehension (PRCA) instrument was used on a sample of 101 students to measure their soft skills. Overall, it was found that students realised the need to enhance their soft skills since it would give them a 'competitive edge' when seeking employment. The PRCA prompted students to reflect on the need to develop team skills and interpersonal skills. No significant difference was found between the attitudes of male and female students. The detailed findings of the study are presented in this chapter.

Key words: soft skills, engineering students, communication skills, public speaking, student feedback.

Introduction

Soft skills play an important role in shaping the personality of an individual by complementing his/her hard skills. Today's employers increasingly demand engineers who not only are competent in their field of specialisation but also possess adequate soft skills (Hillmer, 2007). These include: teamwork capability, communication, professionalism, ethics, problem-solving, lifelong learning and self/time/project/conflict management.

A serious lack in soft skills among engineering graduates has been noted by employers as well as educators (e.g. Zaharim et al., 2009). Research suggested that there is a shortfall in important skills being developed among university students such as: communication, decision making, problem solving, leadership, emotional intelligence and social ethics (Nair et al., 2009). Communication skills have been cited most frequently, followed by knowledge of business or project management. For example, recent concern in this regard was expressed by the British Association of Graduate Recruiters (BAGR), which reported that:

> Employers say many graduates lack 'soft skills', such as team working.

BAGR went on to explain that candidates are normally proficient academically, but lack soft skills, such as communication, as well as verbal and numerical reasoning (British Association of Graduate Recruiters, 2007). Pauw et al. (2006) supported this claim by arguing that poor communication skills tend to make a negative impression on employers during recruitment and may exclude a graduate (with good technical skills) from being selected for employment. The medium of communication in Indian higher education institutions and industries is English. Many Indian graduates, for whom English is their second language, may be perceived to express their ideas very poorly in English.

Thus, for many Indian engineering students it is important to acquire adequate soft skills, including communication skills (in English), that will make them more employable, in addition to the technical skills that they acquired in higher education. This presents a challenge for engineering education to develop lifelong learners (their students) and also to bridge the perceived gap between the changing engineering practices, employers' expectations and the engineering curricula.

There are several observations relevant to the perceived lack of soft skills development. First, most first-year male engineering students report a high level of confidence in their own ability in both technical subjects and subjects developing soft skills (Bestergfield-Sacre et al., 2001; Felder, 1995; Leslie et al., 1998). Second, these young students are reported to experience difficulty in taking advice from parents and teachers (Lloyd,

1998). It also seems likely that, as teachers, engineers may fail to respond to students' attitudes and may not be equipped to teach topics such as teamwork and presentation skills in an accessible manner (Pulko and Parikh, 2003).

Hard and soft skills

Hard skills are those technical skills that are required for a specific profession (Zaharim et al., 2009). For example, in the case of a mechanical engineer, the hard skills would be the professional's knowledge of and ability with machines; for a software engineer, it would be his/her level of proficiency with a programming language. On the other hand, soft skills are not that easy to define. Soft skills such as leadership, negotiating, listening and conflict mediating are as important as hard skills for today's global workforce (Smith, 2010).

The importance of soft skills

There is no doubt that hard skills are essential to perform an engineering job competently and efficiently. A blend of both is needed for professional success as an engineer (Hannon, 2003). Ethaiya and Mangala (2010) argue that soft skills are learned behaviours which enable all students with a strong conceptual and practical framework to build and develop their overall personality and enhance their career prospects (see Figure 4.1). They determine an engineer's attitude towards his/her work, organisation, clients and colleagues. Soft skills are not just limited to the professional workplace, but the need for them also extends to other spheres of life, for example society and family. These skills are not just about communicating, but include the ability to manage stress, to organise and to provide solutions. Most of the time the importance of soft skills is ignored, and is not given adequate attention by engineers while developing their technical or hard skills. Engineering education in India often does not concentrate on these essential soft skills. The curriculum tends to ignore the fact that an engineer will be working in a team, reporting to someone else, writing reports, dealing with work pressures, giving presentations, etc. In all such situations, along with technical skills, an individual's experience and maturity tend to play an important role (Hannon, 2003; Ethaiya and Mangala, 2010). Although there is consistent evidence in the literature of the need for both soft and

Figure 4.1 Various roles played by soft skills

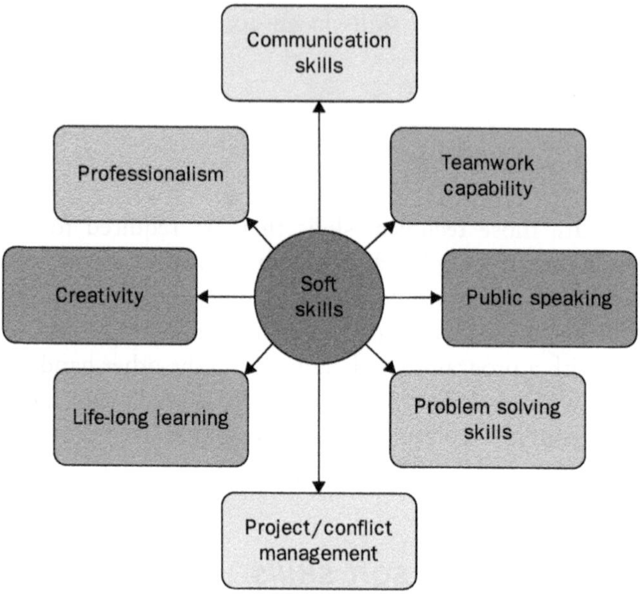

hard skills in engineering education, there is little in terms of understanding the student experience in terms of the learning skills they believe they need for their future employment in India. This chapter presents a case study of how the Jaipur Engineering College and Reseach Centre (JECRC) attempted to explore student perspectives in this area.

Communication skills

In India, an engineering graduate's success in on-campus recruitment is mainly based on their demonstration of effective communication skills. This takes many forms, such as: speaking, writing and listening, though its purpose is always to convey a message to the recruiters who are interviewing the students. Most students are not 'industry ready' because they lack communication skills (Infosys, 2008). Communication skills are an essential component of an engineer's education, and establishing competence in this field should be considered to be a fundamental component of engineering education (Patil and Riemer, 2004).

The parties in the recruiting process may have different wants, needs and attitudes. These can present barriers and thus block success for the student

(Bovee and Thill, 2010). Martin et al. (2005) also reported that graduates felt that, overall, they were well prepared for work in industry. They perceived their strengths to be their technical background, problem-solving skills, formal communication skills and lifelong learning abilities. Areas of weakness were also identified. These included: working in multi-disciplinary teams, leadership, practical preparation and management skills. Students should be trained in the skills which recruiters look for during on-campus recruitment. Based on the assumption that if students are helped to overcome their communication apprehension (communication apprehension is defined as an individual's level of fear or anxiety associated with either real or anticipated communication with another person or persons), they will then be able to develop their communication skills better and thus efforts by trainers to develop the students' employability will be more effective.

A range of methods can be used to improve the communication skills of engineering undergraduates. These methods can be grouped into five main types (Martin et al., 2005):

- writing
- speech
- symbolic gestures
- visual images
- multimedia

Teamwork capability

The soft skills cannot replace the deep technical knowledge of an engineering specialist. However, they are essential if the engineer is to carry out his/her work effectively (Lennartsson and Davidson, 1997). One of such skills is team working. How can we provide feedback that is helpful to team members? How can we develop students' confidence in their teamworking skills?

Research conducted over the past 30 years may offer some answers to these questions. It suggests that the development of any skill is best facilitated by allowing students to practise it and not just by talking about it or demonstrating what to do (Woods, 1993, 1995; Bandura, 1982). There is a need for effective leadership. It is argued that to be an effective team leader, an individual needs to:

- Ensure that everyone on the team is working towards agreed, shared objectives.

- Be able to criticise constructively.
- Continuously monitor the activities of the team members by obtaining effective feedback.
- Constantly encourage and organise the generation of new ideas within the team.
- Always insist on the highest standards of execution from team members.
- Develop the individual and collective skills of the team, and seek to strengthen them by training and recruitment.

The following section examines student perceptions on their soft skills and the need for developing these in a study conducted by the JECRC.

Student feedback

Generally, feedback from students can be used as a measure of teaching effectiveness and also for other aspects of student experience. Using regular student feedback can encourage, enthuse and enhance teaching practice as well as students' educational experiences. When collecting feedback, it is important to consider areas of professional competence (knowledge, skills, communication and attitudes) and also to seek feedback from multiple stakeholders on multiple occasions. This study demonstrates that feedback and evaluation from students is highly valued as a way to enhance the educational outcomes for both teachers and learners.

This study has utilised a range of survey techniques to collect student feedback. These included:

- posted questionnaires
- one-to-one interviews
- focus groups
- student panels

Ensuring good feedback requires that the survey technique is properly used and well suited to the task. It should also allow input from the students where they should be given the chance to comment on the fairness of the feedback and to provide explanations. At the same time it should also involve attentive listening and focus on a positive attitude.

The study found questionnaires to be the most usual and effective type of feedback.

Background to the case study

JECRC, located in the state of Rajasthan, is one of the leading technical institutions in North India. It offers a four-year undergraduate degree programme (Bachelor of Engineering) in the following disciplines:

- Electronics and communication engineering
- Electrical engineering
- Computer science and engineering
- Biotechnology engineering
- Information technology (IT)
- Mechanical engineering

JECRC nurtures the essence of growth in education and its holistic approach focuses on the overall development of its students.

The research bureau of the Associated Chambers of Commerce and Industry of India (ASSOCHAM) identified that industry considers oral communication to be the most important soft skill for future careers, followed by written communication and team work. To evaluate the importance and potential of soft skills, ASSOCHAM conducted a survey and published their findings in a PRCA (Berger et al., 1984; Shumaker and Rodebaugh, 2009) (see Appendices 4.1 and 4.2 for details of this questionnaire). This questionnaire forms part of the study reported in this chapter.

Supporting the need for the study conducted at JECRC described in this chapter was the outcome of mock interviews of the on-campus recruitment conducted by the present authors. The interviews revealed some interesting trends. For example, a student's academic performance could not be directly correlated with their technical skills (hard skills) and practical skills for a number of the students. This was also true for the correlation between teamwork, oral and written communication skills and performance in on-campus recruitment. The literature suggests that hard skills contribute to only 15 per cent of one's performance, while the remaining 85 per cent is made up of soft skills (Crosbie, 2005; Ethaiya and Mangala, 2010). Most employers want to hire, retain and promote employees who are dependable, resourceful, ethical, self-directed and communicate effectively.

Method utilised to investigate the level of soft skills in JECRC students

On the basis of enthusiasm and willingness, 101 undergraduate engineering students from the fourth year of the programme were randomly selected to complete two questionnaires. The first questionnaire required the students to comment on the perceived role of soft skills in their career growth and also in their life in general. The second was based on a PRCA which was used to measure communication apprehension. The performance of each student in soft skills, namely: group discussion, public speaking, interpersonal and communication skills were self-evaluated. The students were given a brief explanation of the aims of the study, but were not told that the results would also be analysed for gender bias. The students were asked to grade each item using a five-point Likert scale (5 = strongly agree, 4 = agree, 3 = neutral, 2 = disagree, 1 = strongly disagree).

Soft skills in career growth: student and facilitator perspectives on enhancing soft skills

In this survey, students were asked to rank the following soft skills in ascending order according to their preference:

- oral communication
- written communication
- teamwork
- research/report writing skills
- analytical ability
- decision making
- fluency in second language
- information management
- time management
- leadership

- networking
- presentation skills
- multi-tasking
- planning
- IT skills

The latter part of the PRCA questionnaire (see Appendix 4.1) comprised questions which related to the students' opinions on the impact of soft skills on their personal and career development. A majority of the students believed that soft skills impacted on their personal and career development.

Findings

The feedback from students suggests that most of them believed that they were weak in report/research skills, information management, fluency in second language, networking, IT skills and the ability to multi-task. Most of them believed that they were good at team work, analytical ability, decision making, oral and written communication, leadership and planning (see Figure 4.2). While 25 per cent of the students perceived and ranked their oral and written communication as weak, 18 per cent evaluated their analytical ability and decision making as weak.

Figure 4.2   Student ranking of their soft skills

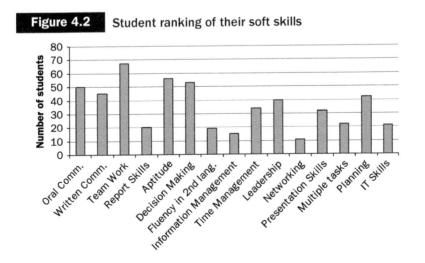

Students' suggestions for enhancing their soft skills

A range of comments were received from students suggesting how their soft skills should be enhanced. Student responses are discussed below, according to individual questions in the survey.

1. Can you suggest in what ways would you improve your communication skills?

A significant number of students wanted to improve their fluency in English and their hesitation in speaking. Some students suggested that participating in group discussions and conducting mock interviews would be helpful. Other suggestions included: reading newspapers and novels, listening to the news in English and presenting views in front of an audience. Some also believed that enhancing knowledge and awareness might boost their self-confidence and reduce their reluctance to speak.

2. Can you suggest some ways in which to increase your confidence?

One student, for example, suggested that participating and interacting in various events, organising events and taking part in mock group discussions and interviews can help to increasing confidence. Some others suggested becoming involved in leadership work. Students also emphasised the value of writing research papers and presenting them at conferences. Another suggestion was to create an encouraging and healthy competitive environment that could boost confidence levels. The most common suggestion was to create opportunities for frequent discussions of common topics in groups.

Communication apprehension

The assessment of engineering students' communication apprehension was based on the PRCA questionnaire. The questionnaire looked at the following four areas:

- group discussion
- meetings
- interpersonal communication
- public speaking

The students were required to indicate the degree to which each statement applied to them (see Appendix 4.2 for the detailed questions).

The overall PRCA scores were in the range 24–120. Less than 50 meant low communication apprehension and above 70 was interpreted as high communication apprehension.

Figure 4.3 shows the overall communication apprehension of the 101 students who completed the PRCA questionnaire. Their mean value of communication apprehension was 64.0. The highest score was 91 and the lowest 35. The standard deviation was 12.3.

Figure 4.4 reports the PRCA scores of the sample group. Only four out of 101 students (3.9 per cent) reported low communication apprehension;

Figure 4.3 Overall communication apprehension of respondents to the PRCA questionnaire

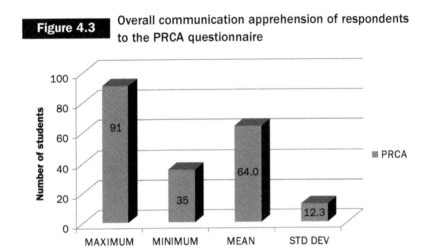

Figure 4.4 PRCA scores of the sample group

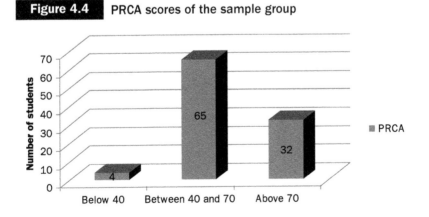

65 students (64.9 per cent) reported a medium level of communication apprehension and 32 students (31.9 per cent) reported a high level of communication apprehension. The results suggest that more than half of the sample group had high levels of communication apprehension.

The mean value of communication apprehension in public speaking was also calculated in the following four areas: group discussion, meetings, interpersonal and public speaking (see Figure 4.5). The results in this instance suggested that students had high levels of apprehension about public speaking thus indicating that students needed more training in public speaking (oral presentation) skills.

Apprehension in group discussion

The analysis related to the group discussion showed that 21 students reported low levels of apprehension about communicating, 46 reported a medium level of apprehension and 34 were very apprehensive in this area (see Figure 4.6).

Apprehension in meetings

The analysis of the group discussion dimension showed that only 25 students had a low level of communication apprehension, 41 had a medium level, and 35 had a high level of apprehension about meetings (see Figure 4.7).

Figure 4.5 Mean value of communication apprehension

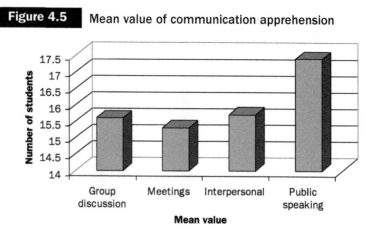

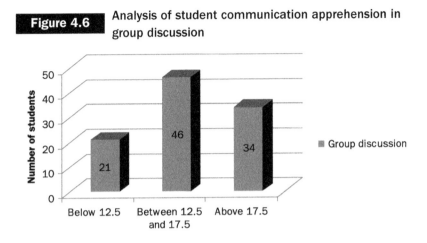

Figure 4.6 Analysis of student communication apprehension in group discussion

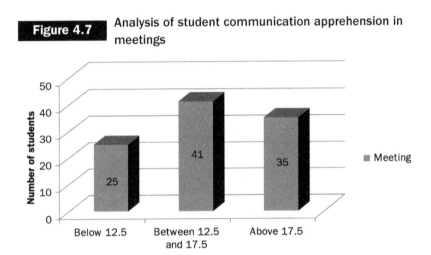

Figure 4.7 Analysis of student communication apprehension in meetings

Apprehension in interpersonal communication

In terms of interpersonal skills, the results showed that 19 students had low communication apprehension, 50 had a medium level of apprehension, and 32 students were very apprehensive about interpersonal communication (see Figure 4.8).

For the domain of public speaking, six students indicated a low level of apprehension, 48 students indicated a medium level and 47 students a high level of apprehension (see Figure 4.9).

Figure 4.8 Analysis of student communication apprehension in interpersonal communication

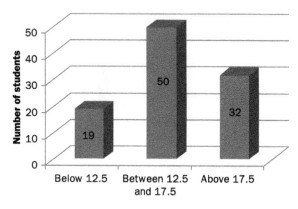

Figure 4.9 Analysis of student communication apprehension in public speaking

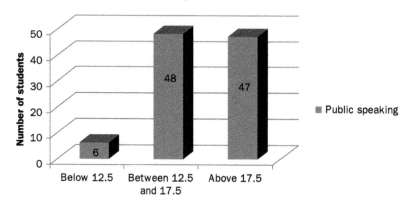

PRCA on gender basis

Of the 101 students who participated in the questionnaire, 73 were male and 28 female. Statistical analysis was carried to investigate the difference between male and female student responses. Two independent t-tests were carried out to compare the differences in a range of parameters, such as, group discussions, meetings, interpersonal situations and public speaking. The tests showed no significant difference between male and female samples over all the parameters. As the sample size varied in both groups, 28 male students were randomly selected, and then compared with 28 female students. This also showed no significance.

Strategies used to enhance soft skills

As a result of this student feedback, a 20-day soft skills programme was implemented to overcome the communication apprehension found among most of the fourth-year engineering students at JECRC. The programme incorporated the following modules:

1. Business communication skills:
   - English language enhancement
   - the art of communication
2. Intrapersonal and interpersonal relationship skills:
   - intrapersonal relationships
   - interpersonal relationships: how to be an effective team player
3. Campus to company:
   - corporate dressing
   - corporate grooming
   - business etiquette
   - communication media etiquette
4. Group discussions, interviews and presentations:
   - group discussions
   - interviews
   - presentations
5. Development of entrepreneurial skills:
   - goal setting
   - entrepreneurial skills: awareness and development.

Each module consisted of 16-hour blocks, including an interactive classroom session. Students were also given tasks and assignments for routine evaluation.

In the group discussion classes, students were divided into groups, and several mock group discussion sessions were conducted. The students were asked to give a three-minute talk as a diagnostic test on a general topic and were marked on the following criteria:

- confidence level
- effective communication
- appropriate body language
- comfort in handling questions
- self-assessment

At the end of their presentation, the students were also asked about their feelings before and while giving their oral presentations. More than 60 per cent of the students scored less than 50 per cent and they attributed their poor performance to a lack of exposure and practice. 40 per cent of the students said that it was their first experience of speaking in front of an audience.

In the PRCA analysis, the mean value of communication apprehension in public speaking was the highest in the categories group discussion, meetings and interpersonal communication. It was found that, in most cases, the students' PRCA in public speaking and their performance in the diagnostic speaking test matched with each other.

Special emphasis was given to English language enhancement. Facilitators were carefully chosen to act as counsellors, communication skill consultants and soft-skill trainers. Students were given training in both oral and written communication. Focus was on the places where students were weak, and hence personal or individual training needed to be improved. To overcome the communication apprehension, three stages were proposed:

1. group sharing
2. one-to-one sessions
3. learner-centred training

Interactive sessions by randomly selecting the groups were also proposed.

Students were only trained only in group discussions and public speaking. They were asked to actively participate in group discussions and speak on informal topics in front of the class. The outcome was very positive and encouraging. The students also felt that mock interview sessions were very effective in increasing their confidence and overcoming their fear of speaking in front of an interview board.

Conclusions

The importance of soft skills when seeking employment has significantly increased in recent times. It is essential to acquire these skills along with technical skills. This chapter discussed a survey on student perceptions of the importance of soft skills for their future employment. It outlined how soft skills can complement hard skills. The main objective of the study described here was to identify the deficiencies in soft skill development among engineering students and propose ways of addressing these deficiencies. This chapter has further highlighted the value of student PRCAs and their suggestions of how these should be addressed. Analysis of data showed no significant difference between male and female students in a range of parameters of group discussion, meetings, interpersonal, public speaking and PRCA.

However, the approach utilised by JECRC had some limitations. The authors found that soft skill enhancement modules were not popular with the engineering students. Some students were, for example, reluctant to attend and others claimed that some modules were irrelevant to them or that they had already covered the topics. The data confirmed that most of these students were male and they exhibited more confidence in their own technical and soft skills. Student feedback helped the authors to explore and build awareness among students of the impact of soft skills in their career as well as personal development.

Acknowledgements

The authors would like to thank the management of JECRC for their support and the final-year students who, directly or indirectly, helped to provide information and feedback regarding their soft skills.

Appendix 4.1: importance of soft skills from students' perspective

1. Please rank the soft skill in ascending order of your preference:

Soft skill	Rank/preference
Oral communication	
Written communication	
Team work	
Research/report skill	
Analytical ability	
Decision-making	
Fluency in second language	
Information management	
Time management	
Leadership	
Networking	
Presentation skill	
Coping with multiple tasks	
Planning	
IT skill	

Which of the above skills are perceived to be important but are scored low by you during higher education?

2. Soft skills should be enhanced/learned during higher education.

    Yes ☐   No ☐

3. Individual personality plays an important role in acquiring soft skills.

    Yes ☐   No ☐

4. Soft skills are decisive during recruitment.

    Yes ☐   No ☐

5. Certification is not an end in itself. Soft skills do matter.

    Yes ☐   No ☐

6. Soft skills help to advance in your career.

    Yes ☐   No ☐

7. Soft skills empower you and create opportunities.

    Yes ☐   No ☐

8. Soft skills not only improve your career, they also offer personal growth.

    Yes ☐   No ☐

9. Should you work on communication and leadership?

Yes ☐ No ☐

10. Soft skills help you grow beyond money motivation.

Yes ☐ No ☐

Suggestions

11. Can you suggest some improvements in yourself to increase your communication skills?

....

....

....

....

12. Can you suggest some ways to bring some improvement in yourself to increase your personality and confidence level?

....

....

....

....

Appendix 4.2: personal report of communication apprehension

Please indicate the degree to which each statement applies to you by marking whether you: strongly disagree = 1; disagree = 2; are neutral = 3; agree = 4; strongly agree = 5

No.	Statement	Score
1.	I dislike participating in group discussions	☐
2.	Generally, I am comfortable while participating in group discussions	☐
3.	I am tense and nervous while participating in group discussions	☐
4.	I like to get involved in group discussions	☐
5.	Engaging in a group discussion with new people makes me tense and nervous	☐
6.	I am calm and relaxed while participating in group discussions	☐
7.	Generally, I am nervous when I have to participate in a meeting	☐
8.	Usually, I am comfortable when I have to participate in a meeting	☐
9.	I am calm and relaxed when I called upon to express an opinion at a meeting	☐
10.	I am afraid to express myself at meetings	☐
11.	Communicating at meetings usually makes me uncomfortable	☐
12.	I am relaxed when answering questions at a meeting	☐
13.	While participating in a conversation with a new acquaintance, I feel very nervous	☐
14.	I have no fear of speaking up in conversations	☐
15.	Ordinarily I am very tense and nervous in conversations	☐
16.	Ordinarily I am calm and relaxed in conversations	☐
17.	When conversing with a new acquaintance, I feel very relaxed	☐
18.	I'm afraid to speak up in conversations	☐
19.	I have no fear of giving a speech	☐
20.	Certain parts of my body feel very tense and rigid while giving a speech	☐
21.	I feel relaxed while giving a speech	☐
22.	My thoughts become confused and jumbled when I am giving a speech	☐

No.	Statement	Score
23.	I face the prospect of giving a speech with confidence	☐
24.	While giving a speech, I get so nervous I forget facts I really know	☐

References

Bandura, A. (1982) 'Self-efficacy mechanism in human agency'. *American Psychologist*, 37(2), 122–47.

Berger, B.A., Richmond, V., McCroskey, J.C. and Baldwin, H.J. (1984) 'Reducing communication apprehension: is there a better way?' *American Journal of Pharmaceutical Education*, 48 spring. Available online at: *http://www.jamesmccroskey.com/publications/117.pdf* (accessed October 2010).

Besterfield-Sacre, M., Moreno, M., Shuman, L.J. and Atman, C.J. (2001) 'Gender and ethnicity differences in freshmen engineering student attitudes: a cross-institutional study'. *Journal of Engineering Education*, 90(4), 477–89.

Bovee, C.L. and Thill, J.V. (2010). *Business Communication Today* (9th edn.). Upper Saddle River, NJ: Prentice Hall by Pearson Education Inc.

British Association of Graduate Recruiters (2007) Available online at: *http://www.agr.org.uk/* (accessed October 2010).

Crosbie, R. (2005) 'Learning the soft skills of leadership'. *Industrial and Commercial Training*, 37(1), 45–51.

Ethaiya, R. and Mangala, S. (2010) 'Need and importance of soft skills in students'. *Journal of Literature, Culture and Media Studies*, 2, 1–6.

Felder, R.M. (1995) 'A longitudinal study of engineering student performance and retention. IV: instructional methods and student responses to them'. *Journal of Engineering Education*, 84(4), 361–67.

Hannon, K. (2003) 'Educators are struggling to prepare well rounded engineers for today's workplace'. *AEE prism online*. Available online at: *http://www.prism-magazine.org/mayjune03/* (accessed October 2010).

Hillmer, G. (2007) 'Social and soft skills training concept in engineering education'. Paper presented at *International Conference on Engineering Education (ICEE) 2007*, Coimbra, Portugal, 3–7 September 2007.

Infoys (2008) Available online at: *http://campusconnect.infosys.com/login.aspx* (accessed October 2010).

Lennartsson, B. and Davidson, K. (1997) 'Team understanding capability: the new requirement on higher engineering education'. *Proceeding of the Second International Conference on Teaching Technology at Tertiary Level*. Stockholm, 14–17 June 1997.

Leslie, L.L., McClure, G.T. and Oaxaca, R.L. (1998) 'Women and minorities in science and engineering: a life sequence analysis'. *Journal of Higher Education*, 69(3), 239–76.

Lloyd, T. (1998) *Young Men, the Job Market and Gender Work*. York, UK: Joseph Rowntree Foundation.

Martin, R., Maytham, B., Case, J. and Fraser, D. (2005) 'Engineering graduates perceptions of how well they were prepared for work in industry'. *European Journal of Engineering Education*, 30(2), 167–80.

Nair, C.S., Patil, A. and Mertova, P. (2009) 'Re-engineering graduate skills – a case study'. *European Journal of Engineering Education*, 34(2), 131–9.

Patil, A.S. and Riemer M.J. (2004) 'English and communication skills curricula in engineering and technology courses in the Indian State of Maharashtra: issues and recommendations'. *Global Journal of Engineering Education*, 8(2), 209–18.

Pauw, K., Bhorat, H., Goga, S., Ncube, L. and van der Westhuizen, C. (2006) *Graduate Unemployment in the Context of Skills Shortages, Education and training: Findings from a Firm Survey*. Development Policy Research Unit, Working Paper 06/115. University of Cape Town, South Africa.

Pulko S.H. and Parikh, S. (2003) 'Teaching 'soft' skills to engineers'. *International Journal of Electrical Engineering Education*, 40(4), 243–54.

Shumaker, E.A. and Rodebaugh T.L. (2009) 'Perfectionism and social anxiety: rethinking the role of high standards'. *Journal of Behavior Therapy and Experimental Psychiatry* 40(3), 423–33.

Smith, A.Z. (2010) 'Assessing educational fundraisers for competence and fit rather than experience: A challenge to conventional hiring practices'. *International Journal of Educational Advancement*, 10, 87–97.

Woods, D.R. (1993) 'Problem solving what doesn't seem to work'. *Journal of College Science Teaching*, 23(1), 57–8.

Woods, D.R. (1995) *Problem-based Learning: Resources to Gain the Most from PBL*. Hamilton, OT, Canada: McMaster University Bookstore.

Zaharim, A., Yusoff, Y.M., Omar, M.Z., Mohamed, A., Muhamad, N. and Mustapha, R. (2009) 'Perceptions and expectation toward engineering graduates by employers: a Malaysian study case'. *WSEAS Transactions on Advances in Engineering Education*, 9(6), 296–305.

5

Engineering programmes in Thailand: enhancing the quality of student feedback

Kalayanee Jitgarun, Ake Chaisawadi, Pinit Kumhom and Suthee Ploisawaschai

Abstract: Student feedback can be a powerful tool in improving the quality of learning and teaching in higher education. For course evaluation, students are generally required to provide feedback about an instructor/facilitator's performance (student–teacher feedback), in this case on a rating scale questionnaire. Often such feedback only reflects student satisfaction instead of enhancing learning and teaching. A more meaningful form of feedback that may be used to inform effective learning is one-to-one feedback (either student–teacher or teacher–student feedback). Engaging in hands-on practical work and project activities with students can also generate useful feedback for instructors/teachers. Currently, general education courses at many Thai universities aim to equip students with skills related to the provision of effective feedback such as: asking the right questions and non-judgmental criticism. It is strongly believed that, in the near future, social media will play an important role in 360-degree feedback which may, for example, provide further support to the students to foster their learning. Through a case study undertaken at King Mongkut's University of Technology Thonburi, this chapter explores how student feedback in engineering programmes in Thailand can enhance the quality of student learning outcomes.

Key words: student feedback, engineering education, reciprocal teaching and learning, problem-based learning, project-based learning, team-based learning.

Introduction

In the 1960s there were eight Thai public universities offering undergraduate engineering programmes. By the 1980s the shortage of engineers was of such concern that new public universities and several private colleges were set up and upgraded to engineering and technology universities. All of these institutions operate under the guidelines of the Office of the Higher Education Commission[1] in Thailand, which limits the total number of credit points for all Bachelor degree programmes to 120–150, including a minimum of 30 credit points for general studies and free electives such as: physical sciences, mathematics, English language, social sciences and humanities. Bachelor degree programmes in engineering generally reach the maximum level of 140–50 credit points over eight semesters or four academic years (Wibulswas and Tantarana, 2010). The process is also overseen by the Engineering Profession Control Committee (EPCC), a professional body which issues licenses to practice for civil, electrical, industrial, mechanical and mining engineers.

There are three ways in which a Bachelor's degree in engineering education can be obtained in Thailand. The first approach puts all the first-year engineering students together to undertake common studies, and then separates them into various disciplines from the second year onwards. The second approach separates students into their respective disciplines as soon as they are admitted to their first year. The third approach takes vocational students, who possess a diploma in vocational education, and places them into appropriate disciplines. Relevant subjects from their prior vocational education may be credited to the programme.

Regarding the quality assurance of higher education in Thailand, until relatively recently, the public universities have not been systematically monitored. This happened because the public universities have been fully funded by the government and they have had a system of selecting high-quality students. Consequently, little had been done to ensure the quality of the teaching and learning. However, in order to effectively implement quality assurance in higher education, the Thai government realised that there was a need for an organisation that would establish the criteria and methods to be used. Thailand's National Education Act (Ministry of Education, 1999) proposed establishing such a body to regulate quality in higher education:

> An Office for National Education Standards and Quality Assessment shall be established as a public organization, responsible for development of criteria and methods of external evaluation,

conducting evaluation of educational achievement in order to assess the quality of institutions, bearing in mind the objectives and principles and guidelines for each level of education as stipulated in this Act. . . . (Ministry of Education, 1999: 23)

Apart from the need for an external authority, it is an important part of the internal quality assurance of higher education institutions to ensure that the necessary conditions are in place to develop different types of learning outcomes for the students. According to the Ministry of University Affairs (2006), these are as follows:

- ethical and moral development;
- acquisition of knowledge;
- cognitive skills;
- interpersonal skills and personal responsibility;
- analytical and communication skills.

This chapter explores how student feedback in engineering programmes in Thailand can enhance these learning outcomes and the general development of the students. It examines the types of student feedback in engineering programmes in Thailand that can be generated in classrooms, learning activities, laboratories, workshops and projects.

Requirements of engineering programmes

At present, engineers around the world are required to possess multi-disciplinary knowledge and skills to perform professional work. Employers no longer have the capacity to train their employees on-the-job or through an apprenticeship, to equip them with the required knowledge of mathematics, physical sciences and management techniques. Consequently, more demand is placed on engineering educators to equip students with the required knowledge and skills, as indicated in the following statement by the Accreditation Board for Engineering and Technology (ABET) in Thailand:

> . . . to apply knowledge of mathematics, science, and engineering . . . to design and conduct experiments, as well as to analyse and interpret data . . . to design a system, component, or process to meet desired needs within realistic constraints . . . to function on multidisciplinary

teams . . . to identify, formulate, and solve engineering problems . . . to communicate effectively . . . to understand the impact of engineering solutions in a global, economic, environmental, and societal context . . . to engage in lifelong learning . . . a knowledge of contemporary issues . . . to use the techniques, skills, and modern engineering tools necessary for engineering practice. (ABET Engineering Accreditation Commission, 2009: 3)

Collecting student feedback, together with employer and other stakeholder feedback may significantly assist the development of engineering programmes that would help to develop the wide range of skills that are now required of engineering graduates.

Types of student feedback in engineering programmes in Thailand

It is often argued that the value of student feedback in enhancing the quality of instruction is not clear-cut. Although, in many instances students' evaluation of the instruction could lead to improvements in the effectiveness of the teaching (Murray, 1997), both the extent to which student ratings are valid (Kulik, 2001) and which strategies should be utilised in student feedback (Penny and Coe, 2004) have been questioned.

Feedback from the student to the teacher can be a powerful tool to improve the quality of teaching in engineering in Thailand, because teaching and learning can be synchronised when the teachers are open to receiving feedback from students that indicates their understanding, misconceptions, errors and engagement. In engineering programmes, feedback from students may take a number of forms including achievement levels in assignments and examinations (such as quizzes and formal mid- and final term papers) as well as forming part of the regular process of course evaluation. For course evaluation, students are generally required to provide feedback about an instructor/teacher's performance (student–teacher feedback). Typically, such questions require students to consider aspects of the teaching process. These may include punctuality, and the quality of teaching in lecture and laboratory sessions. Often such feedback only reflects student satisfaction instead of improvement. Engaging in hands-on practical work and project activities with students can also generate useful feedback for the instructors/teachers.

King Mongkut's University of Technology Thonburi (KMUTT): a case study

According to the goals of the engineering programmes at KMUTT, the superficial feedback produced from ticking boxes on marking sheets is considered ineffective and non-dynamic because it is impersonal and does not allow for student comments or deeper reflection. Despite the enormous contribution of time and effort made by instructors, both the process and value of collecting feedback from engineering students remains problematic.

Currently, general courses at KMUTT, as well as at many other universities in Thailand, aim to equip students with broad skills, such as: asking the right questions and non-judgmental criticism which they may utilise in effective feedback provision. It is strongly believed that in the near future social media will play an important role in 360-degree feedback which may reinforce the importance of their voice to the students.

In order for Thai engineering students, especially at those KMUTT, to acquire knowledge and skills and keep abreast of international trends, it is understood that a variety of different teaching and learning environments including classroom learning, seminars, workshops and laboratory sessions are necessary. Project work may also be helpful in developing this knowledge and skills. Student feedback thus serves as a valuable tool for lecturers to further enhance their teaching and student learning. Provision of student feedback in different settings and types of courses at KMUTT is discussed in detail in the following sections.

Feedback in the classroom at KMUTT

This section discusses the observations by the lecturers, and student feedback, on a course titled 'General Education' which is intended for first-year undergraduate students including the engineering students, and how such observations and feedback may trigger improvements in teaching practice.

The aim of the general education course is to offer learning skills to first-year undergraduate students in all fields. Each classroom consists of students from different disciplines such as: sciences, engineering and industrial education. This discussion focuses on the international curriculum in which only engineering students were enrolled. The observations by lecturers noted various forms of feedback by students, and these are detailed later in this section.

In this course, all the lecturers wrote down their observations and feedback from their students each time they attended the class. This was invaluable for the present study. Lecturers also asserted that writing a teaching log after each class has given them an opportunity to reflect on their teaching and that the feedback from students gave them a reference point for better teaching in subsequent classes and semesters. Additionally, students provide their overall feedback to the lecturers at the end of the course in the form of an online questionnaire. However, the authors argue that continuous feedback is also valuable. Below is an outline of significant points taken from the teaching logs regarding student in-class feedback and some implications for future classes.

Because this course entailed both lectures and activities, students encountered two styles of learning in the same course: individual for the first hour and group work for the remaining two hours. At the end of each class, students were asked to recall what they had learned through either a learning memo for online submission or a presentation in class. Feedback from the majority of students suggests that students found the presentation to be a more effective way to recall and express their understanding. Later in the course, at the end of each class a few students (different ones each time) were assigned to give a presentation on what they had learnt and this enabled learning and sharing among students who were keen on the subject and raised interesting points.

It is important to note that oral presentation is an essential skill which students need to practice, apart from writing assignments, because it allows them to improve their confidence and to receive comments from their peers. One advantage of presentation is that instructors can check students' understanding immediately. The course activities dealt with 'fixed' and 'growth' mindsets. Students intended to consider the 'growth' mindset as a 'good' thinking concept, whereas the 'fixed' mindset was seen as 'bad' (although this might have been a misconception). The main point that students should have gained from this exercise was the recognition of which aspects of their thinking fall into the growth mindset and which fall into the fixed one. The aim was to develop the students' capacity to switch between the two mindsets, depending on the situation as fixed mindset might be appropriate for some risky situations. In their final evaluation the students expressed appreciation for the clarification given after presentations during the course.

Another type of feedback which is essential for the improvement of teaching quality comes from reflection by the students. Reflection is also a form of feedback which allows students to deal with the activities in class. Given that the topic of plagiarism provided students with an opportunity to think about their past experiences with plagiarism, they could share their opinions and understanding along with the implications of such actions until they could recognise the risks and appreciate the work of other scholars by citing their sources. Hence, reflection by students encouraged lecturers to cover their topic in more depth. For instance, students' lack of understanding of how to avoid plagiarism encouraged lecturers to teach students how to cite properly.

It may be argued that each type of feedback in the course could be seen as a way of enhancing self-regulated learning in students. According to Zimmerman (1990), self-regulated learning has three features: observing the activities, judging the performance and improving the performance. The course described here offered tasks in which students could build self-regulated learning habits. According to Butler and Winne (1995), feedback is considered to be the most important factor in building up self-regulated learning for learners because feedback is associated with cognitive engagement with tasks. When students are involved in an activity, they tend to have internal and external feedback through their self-monitoring, collaboration with their peers, responses from lecturers and presentation before the end of the class. At the same time, not only the students but also the lecturers adopt self-regulated learning in their classroom. It is often the case that the students are encouraged to be self-regulated, but only a few of them will focus on the feedback and self-regulation by the lecturers (Eekelen et al., 2005).

Seminars at KMUTT

Genalo et al. (2004) argued that classroom instruction is frequently centred on delivering the content to students instead of facilitating student inquiry during the learning process. Holzer (1994) pointed out that student views can be elicited through open-ended questions and encouraged by non-judgmental feedback. In contrast, narrow single-answer (right/wrong) questions tend to discourage risk taking and creativity in students (Brooks and Brooks, 1993). These are some of the reasons why many engineering programmes at KMUTT have provided students with seminar courses. One of the goals of these courses is to acquaint students with a research approach to solving a problem. One approach used at KMUTT has been that a group of students is guided through a project (which has previously been successfully carried out) and this enables the engineering students to learn research methodology. Students are guided to learn to ask questions based on the research process. The overall goal is to change students' mind-set to be more proactive.

If possible, the process can also be carried out in person in a workshop-like manner, utilising instant interactive feedback. From the perspective of feedback, the main difference between this learning process and a usual classroom or laboratory is that the feedback can be frequent, instant and informal.

Discussion and conclusions

This chapter has argued that although student feedback in Thailand and elsewhere is mainly understood as a satisfaction rating at the end of the course, there are other types of feedback that might be utilised in the classroom, workshop, laboratory or project work. However, the validity of student feedback is often questioned (Kulik, 2001). Thus, it is regarded as a requirement on lecturers and is often not taken seriously. However, as Brooks and Brooks (1993) argue, instructors should seek and value student perspectives as windows to their knowledge and reasoning, subsequently aiding enhancement of the teaching and learning process (Holzer, 1994). Learning is something that students must do and take ownership of, rather than something that is done to them (Savage et al., 2007). Utilising student feedback and making appropriate changes to teaching may more successfully facilitate self-directed learning and thus motivate students more effectively. Thai higher education still has some way to travel, before it reaches this stage. Therefore, this chapter has merely attempted to outline the issues and general situation regarding utilising student feedback in engineering education in Thailand.

Acknowledgements

The authors are particularly grateful to Mr. Kenneth Dun of Uttaradit Rajabhat University, Thailand, for his proofreading and editing of this chapter.

Note

1. The Office of the Higher Education Commission has authority in various areas such as: proposing policies for higher education and standards which correspond to the National Economic and Social Development Plan and the National Education Plan; setting criteria and identifying resources to support higher education; and the development of monitoring and evaluation systems for higher education provision on the basis of academic freedom and the excellence of each individual degree-granting institution; considering and issuing regulations, criteria and official orders as deemed necessary.

References

ABET Accreditation Board of Engineering and Technology (2009) Criteria for accrediting engineering programs: effective for evaluations during the 2010–2011 accreditation cycle. Available online at: *http://www.abet.org/forms.shtml* (accessed December 2010).

Brooks, J.G. and Brooks, M.G. (1993) *The Case for Constructivist Classrooms.* Alexandria, VA, USA: Association for Supervision and Curriculum Development.

Butler, D.L. and Winne, P.H. (1995) 'Feedback and self-regulated learning: a theoretical synthesis'. *Review of Educational Research*, 65(3), 245–81.

Eekelen, I.M., Boshuizen, H.P. and Vermunt, J.D. (2005) 'Self-regulation in higher education teacher learning'. *Higher Education*, 50(3), 447–71.

Genalo, L.J., Schmidt, D.A. and Schiltz, M. (2004) Piaget and engineering education. *Proceedings of the 2004 American Society for Engineering Education Annual Conference and Exposition*, Salt Lake City, UT, USA, 20–23 June 2004.

Holzer, S. (1994) From constructivism to active learning. *The Innovator, 2.* Available online at: *http://www.succeed.ufl.edu/innovators/innovator_2/innovator002.html* (accessed November 2010).

Kulik, J.A. (2001) 'Student ratings: validity, utility, and controversy'. *New Directions for Institutional Research*, 109, 9–25.

Ministry of Education, Thailand (1999) *National Education Act of B.E. 2542 (1999).* Available online at: *http://www.moe.go.th/English/edu-act.htm* (accessed March 2010).

Ministry of University Affairs, Thailand (2006) *National Qualifications Framework for Higher Education in Thailand: Implementation Handbook.* Available online at: *http://www.mua.go.th/users/tqf-hed/news/FilesNews/FilesNews8/NQF-HEd.pdf* (accessed November 2010).

Murray, H.G. (1997) 'Does evaluation of teaching lead to improvement of teaching?' *International Journal for Academic Development*, 2(1), 8–23.

Penny, A.R. and Coe, R. (2004) 'Effectiveness of consultation on student ratings feedback: a meta-analysis'. *Review of Educational Research*, 74(2), 215–53.

Savage, R.N., Chen, K.C. and Vanasaupa, L. (2007) 'Integrating project-based learning throughout the undergraduate engineering curriculum'. *Journal of STEM Education*, 8(3), 1–13.

Wibulswas, P. and Tantarana, S. (2010) Engineering and technology education in Thailand. Available online at: *http://www.transworldeducation.com/articles/thailand2.htm* (accessed November 2010).

Zimmerman, B.J. (1990) 'Self-regulated learning and academic achievement: an overview'. *Educational Psychologist*, 25(1), 3–17.

6

Summative quality assurance systems: not good enough for quality enhancement

Roy Andersson, Anders Ahlberg and Torgny Roxå

Abstract: In this chapter we scrutinise an elaborate institutional quality assurance model, with aspirations to develop a quality culture which aims to improve student learning, in order to discuss general issues of teaching and learning evaluation strategies. Our analysis suggests that summative student evaluations are useful for institutional quality assurance and quality enhancement at study programme level. However, they appear less efficient for quality enhancement at course module or subject discipline level, that is the loci of teaching and learning. To support quality enhancement of teaching and learning, iterative formative evaluation has greater potential and in order to promote an institutional quality culture, summative as well as formative student evaluations need to be in place, discussed, accepted and understood by all legitimate stakeholders (i.e. students, university teachers and institutional representatives).

Key words: teaching and learning, higher education, quality assurance, quality enhancement, summative assessment, formative assessment.

Introduction

In this chapter we use an existing, elaborate, student evaluation system as the background to discussing general issues of evaluation strategies. The setting is the Faculty of Engineering (Swedish acronym LTH), a research-intensive semi-autonomous faculty within Lund University. Lund University, which is one of Scandinavia's largest institutions for education and research, enrols 46,000 students and has 6,000 employees, within which LTH has arround 8,000 students and 1,400 employees (LTH, 2010).

Several research groups at LTH are considered to be of high international standard, and some are world leaders, such as those in nano-technology, combustion physics, mobile communications, automatic control, laser physics and biotechnology (LTH, 2010). In addition, LTH has earned national and international attention for its theoretically underpinned (Barr and Tagg, 1995; Bowden and Marton, 1999; Biggs, 2003; Ramsden, 2005) programme to improve teaching and learning that started in the year 2000. LTH also has a reputation for taking the enhancement of student learning seriously. The core strategy of the programme is to support scholarship in teaching and learning (Boyer, 1990; Trigwell et al., 2000; Kreber, 2002) among all employees by promoting pedagogical dialogues. This is anticipated to lead to innovation, improved teaching, and better student learning: in other words, an emerging quality culture (Andersson, 2010; Harvey and Stensaker, 2008; Mårtensson et al., 2011).

Despite these efforts the LTH congruent learning-centred elaborate student evaluation system (described in detail below) does not seem to entirely match what could be expected of a teaching and learning quality culture. This indicates that making a student evaluation system successful is not an obvious task, but by scrutinising the enforced evaluation strategies at LTH we hope this chapter will help others to improve their strategies – at both institutional and individual university teacher level.

Terminology

Some of the terms we use vary according to the context, so we will clarify what we mean with some key terms:

- *Course or course module:* a course module taught and assessed as a stand-alone unit.
- *Programme:* a degree study programme (Bachelor, Master) consisting of several courses.
- *Summative evaluation:* an evaluation accomplished *after* an activity/course, with the focus on summarising what has happened. It may contain data collected both during and after the activity/course.

- *Formative evaluation:* an evaluation accomplished *during* an activity/ course, with the focus on further enhanced student learning.

- *Reporting evaluation:* an evaluation done primarily to inform other (higher) organisational levels or for specific stakeholders, where documentation is a central part.

- *Operational evaluation:* a set of formative evaluations done primarily to improve the ongoing teaching and learning activity or course. It is performed by the university teacher in collaboration with the students with the focus on making the university teacher aware of what is happening in the ongoing activity/course and thereby be able to perform an 'adaptive regulation' of the teaching.

So, the terms summative and formative, respectively, are merely telling *when* an evaluation is accomplished, and reporting and operational, respectively, relate to *the purpose* of the evaluation.

Summative student evaluations

As a part of the programme to improve teaching and learning, LTH developed a uniform faculty-wide student evaluation system, which has been in place since 2003 (Warfvinge, 2003). This includes a questionnaire (web or paper), compulsory dialogue between major stakeholders (lecturers, programme directors and students) and formal public documentation of both evaluation data and comments from the stakeholders. The questionnaire is based on the theoretically and empirically underpinned course experience questionnaire (CEQ) by Ramsden (2005). Answers to 26 multiple choice questions are scored in five clusters of importance for stimulating a deep approach to learning:

- good teaching;
- clear goals and standards;
- appropriate assessment;
- appropriate workload;
- development of generic skills.

In addition overall student satisfaction is monitored, as well as answers to teacher designed open-ended questions on aspects of the particular course module. There is currently a database of 120,000 filled out CEQ forms used for course module analyses, study programme analyses and other thematic analyses.

The procedure for summative student evaluations involves six steps:

1. Students fill out the CEQ form at the end of each course taken. The form is normally filled out on the web.

2. The evaluation system transforms the data into a working report, including all individual comments from the students, the enrolment figures, and the overall results from the examination.

3. The data within the working report is discussed at a mandatory meeting of the course coordinator (main course teacher), student representatives, and the programme director.

4. The course coordinator, the student representatives and the programme director independently write short comments based on the discussions at the meeting. These comments are included in the final report. This is an important part of the process to ensure the major stakeholders are given the opportunity to include their personal reflection on the evaluation in formal public documentation.

5. Statistically processed data and the major stakeholders' comments from the discussion make up the final report. The individual comments from the students are not part of the final report.

6. The final report is then published on the LTH intranet and sent via e-mail to all students who participated in the course and to the head of department responsible for the course.

In addition, once a year, all programme directors write a summary covering all their courses in an annual report. These reports also include automatically collected data from all the final course reports in the programme (Roxå and Mårtensson, 2010). Steps 1–6 are designed to stimulate improvement of the course module and raise awareness of poor teaching and learning, to avoid them being repeated next time the course module is taught.

In spite of the elaboration of this summative evaluation system and its aspiration of fostering deep approaches to learning, it has not fully created a satisfactory quality culture. There is a lack of engagement by some individual university teachers and programme directors (Borell et al., 2008; Roxå and Mårtensson, 2010). However, a recent study indicates that the CEQ student evaluations work well as a tool for quality assurance (Roxå et al., 2007). But do they contribute to quality enhancement? Here the answer is more complex. They do contribute to enhancement on the programme level, as indicated by the programme directors (Roxå et al., 2007). However, they do not seem to contribute much to the teachers' efforts to improve teaching within individual course modules (Bergström, 2010). In-depth interviews with individual teachers show that they:

- Pay little attention to statistical descriptions presented in summative evaluations.

- Pay selective attention to answers to open-ended questions.

- Prefer to refer back to their personal experience during interactions with their students.

- Do not view the end-of-course reports as adding in any significant way to their personal experience of what has taken place during the course.

Some argue that summative student evaluations, which unavoidably are drained of detailed information, appear useless to teachers who have personal experience of their course and teaching (Roxå et al., 2007; Roxå and Mårtensson, 2010). Therefore summative student evaluations of teaching seem unable to significantly contribute to an emerging quality culture of teaching and learning. The reason for this, we argue, is that what an organisational culture absorbs or not is always related to what its members attach meaning to (Alvesson, 2002; Schein, 2004). Meaningful routines and procedures tend to be stabilised and used, while activities associated with lesser meaning tend to wither away. Thus, summative evaluations will not contribute to a quality culture as long as the individual teachers find no or limited value in them.

The overall conclusion so far is that in order to promote an emerging quality culture, both summative and formative student evaluations have to be applied. If the teachers are to use student evaluations for improvement purposes, these have to add to their personal experience of teaching. To become useful for individual teachers, statistical data from student evaluations therefore have to be reconstructed to match the personal experience of teaching, while formative evaluations are already richer in this respect. For this reconstruction to happen, a personal commitment from the teacher is needed; or an evaluation format which naturally adds to the teacher's personal experience. In the following section we therefore, elaborate on the use of formative student evaluations, e.g. classroom assessment (Angelo and Cross, 1993). We consider that these have the potential to add value to the teachers' personal experiences of teaching.

Based on our findings, we make the following claims about the LTH summative student evaluations system:

- On the institutional or faculty level where we mainly seek quality assurance, the system *works well. In* the LTH case, it is possible to analyse the CEQ database and retain meaningful answers.

- On the programme level, we seek both quality assurance in relation to the faculty level and quality enhancement in relation to the study programme. In some sense the programme level is a reporting level since it reports a programme summary to the faculty level, but it is also an operational level since course evaluations are collected formatively throughout the year and used to improve the programme. On this level the system *works very well*.

- On the course level, we mainly seek quality enhancement to improve the course module. On this level the system *is not good enough*. Although, if we add operational student evaluation, we have shown that the LTH summative student evaluation system in combination with operational evaluation *works very well, if* the individual teacher engages in a reconstruction of his or her experience.

Therefore, the main purpose of a summative reporting student evaluation system is quality assurance – which, of course, is a very important issue for an institution or faculty. Nevertheless, for an individual university teacher, quality enhancement at course module level is generally a higher priority. This shows that, for a system with both assurance and developmental purposes, a two-fold evaluation process must be undertaken by the university teachers:

- To contribute to the quality assurance process of an institution or faculty, a single teacher needs to understand the importance of fulfilling the reporting evaluation even if it is not sufficient for the teacher's own course enhancement purposes.

- The individual teacher also needs to understand that the reporting evaluation is not sufficient for the teacher's own enhancement purposes and that the teacher needs to perform an additional operational evaluation.

Formative operational evaluations

In addition to the summative CEQ system, since 2003 LTH has required that formative evaluations (so-called operational evaluations) are conducted throughout all course modules. There are, however, no requirements on the format, and no top-down control on how this is enforced. These operational evaluations span from formalised, organised classroom assessment techniques to highly informal verbal classroom discussions. The common denominator in all forms is that the university teacher initiates a dialogue with the students where he or she gathers information which is useful for monitoring the students' learning, that is, for immediate improvement of teaching and learning. When these evaluations work well, the students experience a university teacher who cares about their learning and is serious about teaching. In turn, this has the potential to further stimulate intrinsic student motivation and performance (Angelo and Cross, 1993).

Operational evaluation on an institutional level: a pilot study

In conjunction with the then new faculty-wide policy of mandatory operational evaluation of all LTH courses, a pilot study was launched involving operational evaluation of all courses taught at the Department of Electro Science during the autumn of 2002 (Larsson and Ahlberg, 2003, 2004). Classes were typically large (up to 150 students) to intermediate in size (around 30 students), and some 20 academics were involved in teaching. They were introduced to the principles of classroom assessment techniques (Angelo and Cross, 1993; Black and William, 1996) and were urged to find and modify these techniques to suit their classroom situations. An educational developer was available to answer questions or assist in introducing operational evaluation. After a six-month trial period, operational evaluation activities were evaluated. Questionnaires showed that in large classes (>80 students), the positive effects of the operational evaluation procedures that had been introduced were obvious to most students. In smaller classes (<30 students), the outcome was less clear, either due to subtleties in informal classroom communication (not obvious to students), or merely due to the absence of formative evaluation. The study further indicated that operational evaluation did indeed enhance learning, and when encouraged by the institution the university teachers found operational evaluation personally and professionally rewarding, as well as valuable for the students. It was postulated that providing a framework at departmental level effectively supported ongoing operational evaluation regardless of the type of course and the personality of the lecturer. Obvious pitfalls included a teacher's fear of diverting from the original course plan, or not realising that sticking to the plan might impede learning.

Important aspects of a system supporting an emerging quality culture

In the following we indicate important aspects of a functioning evaluation system. The emphasis is on what should be considered at the institutional level and at the level of the individual teacher, respectively.

Aspects to consider at an institutional level

Aspects concerning summative reporting evaluation systems

A consistent and research-based summative reporting evaluation system is an important ingredient in an institutional quality assurance system. However, since its value for the individual teacher is not always obvious, a top-down discursive pressure is needed to maintain the system. Since the individual teacher's participation is necessary, we also need to consider the following points:

- The university teachers need to be personally invited and informed about their importance in making the quality assurance process work by conducting the summative evaluation, even if it is not sufficient for the teachers' own enhancement purposes to improve their teaching.
- Further, just to be informed is not sufficient. The teachers also need to see outcomes from the system to fully understand the value of their input. At LTH, over 120,000 course evaluation forms have been conducted since 2003 and at the institutional level they have been used to show, for example, that 'the faculty's teaching has improved over time' (Almqvist and Larsson, 2008), or that 'rewarded university teachers also receive higher scores in evaluations' (Olsson and Roxå, 2008).
- Student evaluations ought not to be made public, unless the comments or reflections of the university teachers who are involved are included in the official documentation.

Aspects concerning operational evaluation

To advance from a quality assurance system, based on summative/reporting evaluation, to a quality culture, operational evaluation needs to be added as an ingredient. The following points need to be considered in order to increase the chances of making operational evaluation a natural part of a quality culture within an institution:

- Individual university teachers need to understand that a reporting evaluation is not sufficient for course enhancement and that additional operational evaluations are necessary.
- University teachers should be informed about the virtues of operational evaluation, but that is not sufficient. They also need to see that operational evaluations work, which can be done by using operational evaluations regarding their thoughts about study programmes and other aspects during teacher training courses.
- Students are intrinsically motivated if they see that they can influence teaching and learning situations. Therefore, it is important to report back to students on how the information collected from them is utilised.
- Effective communication channels and processes need to be built at the local level (within departments and course teams) to support operational evaluation regardless of the type of course and the personality of the teacher. Institution-wide processes are most likely not as efficient, since operational evaluation is highly context dependent.

Aspects concerning quality culture

Our final general point concerns creating a quality culture:

> Evaluation ought not to be perceived as an isolated phenomenon, instead it should become a part of an interactive environment together with other quality enhancing activities. (Andersson, 2010)

Aspects to consider at the teacher level

A successful operational evaluation consists of three equally important parts:

- the teacher should be clear about the objectives;
- the teacher should set the basic conditions for a dialogue;
- the actual operational activities.

In this section we address our own rather concrete recommendations directly to the individual teacher. Some of these suggestions may seem trivial, but teaching routine and lack of support from colleagues often cause even basic habits of good teaching to deteriorate.

Be clear about your objectives

Decide what you want to achieve with your teaching (lecture, lab, seminar, etc.). If you do not do this, it is hard to evaluate anything because all other activities will be more or less random.

Set the basic conditions for a dialogue

To evaluate the effects on student learning from your teaching you need to set the conditions for a dialogue with the students. This is not something that just happens by itself, the university teacher must actively initiate the dialogue.

Some additional concrete things for university teachers to consider are:

- Make yourself available: be on time, do not hurry away from a teaching session. Normally it is in conjunction with classroom activities that the students dare to approach you.

- Show the students that you invite dialogue. For instance, if you ask a question during a teaching session, do not answer it yourself before you have given the students an opportunity to answer. Try to avoid dialogue-restraining questions like 'Do you understand?' or 'Does anyone have any questions?' The students easily misunderstand these questions and may interpret them as 'Do you understand – or are you so stupid that you don't understand everything after my crystal clear instruction?' Do not point in a military manner with your whole open hand to a student. Do answer the students with respect (trivial is a forbidden word if you want a future dialogue).

- Spend some part of a teaching session discussing a question posed by a student during a break (if you think the question is of general interest). Make sure that you mention that this question came from a student.

- If your operational evaluations do change something, be sure to tell the students so that they realise that their input made the change possible.

Actual operational activities

The actual operational activities can be divided into three areas. The first category includes spontaneous activities:

- Let the students answer questions during teaching sessions by raising hands that you count (or let them raise hands with 0–5 fingers to grade something). This works fine if you want a quick feedback on student activities, even in a lecture with 200+ students.

- Spend part of a teaching session discussing a question posed by a student during a break (if you think the question is of general interest). Make sure that you mention that this question came from a student. (This point has already been noted in the previous section of this chapter.)

- Drop in on your students' lab or exercise sessions even if you are not responsible for teaching the lab. This shows that you are genuinely interested in both the students and their learning.

The second category includes structured activities (normally for a specific need):

- Use forms (of the plus/minus/interesting type) (Angelo and Cross, 1993).

- Have regular meetings with teaching assistants.

- Visit all your lab sessions with the purpose of collecting something specific.

- Give the students a diagnostic mini-test.

- Remind the students, especially just before any scheduled meeting, that they should tell their student representatives what they want them to take to the course leader.

Finally, there are formalised activities:

- Use weekly reports from all teaching assistants. One example could be to use weekly reports on how the students manage to keep up with the intended pace of the course. Then you can identify and talk to those students who are in danger of falling behind before it is too late (Andersson and Roxå, 2000).

- Use weekly lunch meeting with a few student representatives to discuss issues and reflections raised among the students.

- Use several diagnostic tests to measure the students' progress. These can be in different forms: anonymous, peer reviewed etc. (cf. Kihl et al., 2007; Weurlander et al., 2010).

Conclusions

Summative student evaluations of teaching are in most cases not enough to improve university teaching. Summative evaluations, especially if they are part of a wider scheme beyond the control of the individual teacher, are always drained of information in comparison to the actual experience of the teacher during a course. Therefore, these evaluations can only complement the experience, or, something that rarely happens, be used to reconstruct the experience in the light of the new data provided by the students. This requires a special effort by the teacher: something only dedicated teachers are likely to engage in. However, summative evaluations are necessary for successful quality assurance, a fact that individual teachers often need to be reminded of.

We have argued that, in order to build a quality culture, operational evaluations must be a regular part of teaching. We have also pointed out important aspects for institutional leaders and individual teachers to consider during the emergence of a quality culture for teaching and student learning.

References

Almqvist, M. and Larsson, B. (2008) 'Kan man på programnivå vända negativa trender?' [Transl: Is it possible to reverse negative trends on programme level?], *5:e Pedagogiska inspirationskonferensen*. Lund, Sweden: LTH, Lund University.

Alvesson, M. (2002) *Understanding Organizational Culture*. London: Sage.

Andersson, R. (2010) 'Improving teaching – done in a context'. In Maria Lucia Giovannini (ed.) *Learning to Teach in Higher Education: Approaches and Case Studies in Europe*, pp. 57–82. Bologna, Italy: Clueb.

Andersson, R. and Roxå, T. (2000) 'Encouraging students in large classes'. *Proceedings of ACM/SIGCSE Symposium 2000*, pp. 176–79. Austin, TX, 8–12 March 2000.

Angelo, T. and Cross, P. (1993) *Classroom Assessment Techniques*. San Francisco, CA: Jossey-Bass.

Barr, R.B. and Tagg, J. (1995) 'From teaching to learning – a new paradigm for undergraduate education'. *Change*, November/December 1995, 13–25.

Bergström, M. (2010) Personal communication.

Biggs, J. (2003) *Teaching for Quality Learning at University*. Buckingham, UK: Society for Research into Higher Education and Open University Press.

Black, P. and William, D. (1996) 'Meaning and consequences: a basis for distinguishing formative and summative functions of assessment'. *British Educational Research Journal*, 22(5), 537–48.

Borell, J., Andersson, K., Alveteg, M. and Roxå, T. (2008) 'Vad kan vi lära oss efter fem år med CEQ?' [Transl: What can we learn after five years with CEQ?]. 5:e Pedagogiska inspirations-konferensen. Lund, Sweden: LTH, Lund University.

Bowden, J. and Marton, F. (1999) *The University of Learning*. London: Kogan Page.

Boyer, E.L. (1990) *Scholarship Reconsidered. Priorities of the Professoriate.* Princeton, NJ: Carnegie Foundation.

Harvey, L. and Stensaker, B. (2008) 'Quality culture: understandings, boundaries and linkages'. *European Journal of Education*, 43(4), 427–42.

Kihl, M., Andersson, R. and Axelsson, A. (2007) 'Kamratgranskning i stora klasser' [Transl: Peer review in large classes]. Utvecklingskonferens LU. Lund, Sweden: LTH, Lund University.

Kreber, C. (2002) 'Teaching excellence, teaching expertise and the scholarship of teaching'. *Innovative Higher Education*, 27(1), 5–23.

Larsson, B. and Ahlberg, A. (2003) 'Continuous assessment in engineering education: a pilot study'. Pedagogisk inspirationskonferens. Lund, Sweden: LTH, Lund University.

Larsson, B. and Ahlberg, A. (2004) 'Continuous assessment'. In A. Kolmos, O. Vinther, P. Andersson, L. Malmi and M. Fuglem (eds.), *Faculty Development in Nordic Engineering Education.* Aalborg, Denmark: Aalborg University Press.

LTH (2010) 'Om LTH' [Transl: About LTH]. Available online at: *http://www. lth.se/omlth/* (accessed September 2010).

Mårtensson, K., Roxå, T. and Olsson, T. (2011) 'Developing a quality culture through the scholarship of teaching and learning'. *Higher Education Research and Development*, 30(1), 51–62.

Olsson, T. and Roxå, T. (2008) 'Evaluating rewards for excellent teaching – a cultural approach'. *The HERDSA International Conference*, Rotorua, New Zealand, 1–4 July 2008.

Ramsden, P. (2005) *Learning to Teach in Higher Education.* London: RoutledgeFalmer.

Roxå, T., Andersson, R. and Warfvinge, P. (2007) 'Making use of student evaluations of teaching in a "culture of quality"'. *29th Annual European Higher Education Society (EAIR) Forum*, Innsbruck, Austria, 26–29 August 2007.

Roxå, T. and Mårtensson, K. (2010) Improving university teaching through student feedback: a critical investigation'. In S.C. Nair and P. Mertova (eds.) *Student Feedback: The Cornerstone to an Effective Quality Assurance System in Higher Education.* Cambridge, UK: Woodhead Publishing.

Schein, E.H. (2004) *Organizational Culture and Leadership* (3rd edn.). San Francisco, CA: Jossey-Bass.

Trigwell, K., Martin, E., Benjamin, J. and Prosser, M. (2000) 'Scholarship of teaching: a model'. *Higher Education Research and Development*, 19(2), 158–68.

Warfvinge, P. (2003) 'Policy för utvärdering av grundutbildning' [Transl: Policy on evaluation of undergraduate courses]. Lund, Sweden: LTH, Lund.

Weurlander, M., Andersson, R., Axelsson, A., Hult, H. and Wernerson, A. (2010) 'How formative assessment act as a tool for learning – theoretical aspects and practical implications'. *The 7th International Society for the Scholarship of Teaching and Learning Conference incorporating the 18th Improving Student Learning Symposium*, Liverpool, UK, 19–22 October 2010.

7

Engineering programmes in the UK: the student feedback experience

James Williams and David Kane

Abstract: In the UK, collecting feedback from students about their experience of higher education has become one of the key elements in national and institutional quality processes. Engineering students, like those from other disciplines, are encouraged to complete feedback questionnaires at all levels from individual course evaluation forms to institutional-level surveys and the National Student Survey. Engineering suffers more than most from poor recruitment, difficulties in motivating students and higher than average drop-out rates. This chapter explores the ways in which this may be reflected in student experience surveys. A review of student feedback surveys at national and institutional level indicates three key issues. First, engineering students seem to be noticeably poor at responding to student experience surveys. Second, engineering students have been slightly less satisfied with items relating to assessment and feedback than some other disciplines. Third, it is arguable that engineering students are less satisfied with aspects of their experience because of conditions that are local to individual institutions rather than anything specifically relating to the engineering student experience. This indicates that surveys carried out at institutional level provide a degree of detail and depth that is difficult to reflect at the broader, national level.

Key words: UK higher education, National Student Survey, student feedback, engineering education.

Introduction

Recruitment and retention of students in UK engineering faculties have been the focus of noticeable concern at policy and institutional level for some years. A recent report in the magazine *New Civil Engineer* described MPs' shock at the high drop-out rate from engineering programmes in comparison with other faculties (Flynn, 2008). Academic research has indicated that, in many cases, young people are unable to follow or complete study programmes in the subject (Parsons, 2003). This is not a new phenomenon and in the UK, as across much of Europe and North America, engineering faculties suffered from poor recruitment particularly in the late 1990s, though this has reportedly changed in recent years (Confederation of British Industry, 2008). Science, technology, engineering and mathematics subjects remained a concern of the Labour government until its demise in 2010, and efforts were made to encourage young people to follow this route through universities (HM Treasury, 2002).

The motivation of young people taking up university places in engineering has attracted much attention in the academic press. As early as 1997, a small-scale research project by the Centre for Research into Quality[1] explored the motivations of young people to take up engineering at that university and identified a number of key factors, such as: personal interests (21.9 per cent of respondents), school work (15.4 per cent) and work experience (14.8 per cent). Where the focus is specifically on engineering students, it is primarily on ways of motivating them to study the subject, the implication being that they are unusually unmotivated (Souitaris et al., 2007; Mustoe and Croft, 1999). In particular, there is concern that engineering students are lacking in mathematical expertise and that this is particularly problematic (Sazhin, 1998).

Many previous studies that involve the experiences of engineering students in higher education appear to do so incidentally and are randomly selected samples (Felder et al., 2002). One of the few works that focuses on engineering students focuses specifically on exploring the value of using a student feedback questionnaire in improving the student learning experience (Gregory et al., 1994). However, it appears to use engineering students merely as a case study rather than as a case of particular interest. Its conclusion is important, however, in highlighting that feedback questionnaires are valuable in supporting management in improving the student experience. An exploration of the experiences of engineering students is therefore valuable as part of a wider attempt to understand what deters young people from studying this subject.

Engineering student feedback is usually only published as a result of student feedback surveys, either at the national level, through such instruments as the National Student Survey (NSS)[2] or at institutional level, through the many student satisfaction surveys that have been carried out at universities as part of internal quality enhancement processes. Institutional surveys tend to be much more in-depth in their treatment of the experiences of their own students than the NSS which, by its very nature, is limited in scope. Historically, institutional surveys often allow a fuller analysis of student experience by a range of demographics, including discipline.

Student feedback and existing data sources

Much has been written on the importance and uses of student feedback (Richardson, 2005; Williams and Cappuccini-Ansfield, 2007; Harvey, 2003; Williams and Brennan, 2003), so it is unnecessary to rehearse the arguments here. However, it is important to note that student feedback has become accepted as one of the standard pillars of the UK's quality assurance and enhancement systems and the importance of 'listening to the student voice' is widely acknowledged (Symons, 2006). Since the development of the NSS in 2005, as a result of the Cooke Report (Higher Education Funding Council for England, 2002) which explored the information requirements in higher education, data on the broad experience of students across the higher education sector has been collected nationally. A short survey, based on a range of items relating to the core aspects of the student experience, the NSS has proved controversial. Despite this interest in the collection and use of student feedback, surprisingly little research beyond institutional reports and national surveys such as the NSS, has been undertaken that explores the way in which students from different disciplines respond to student experience surveys.

The NSS, although influential, is limited in what it tells us about the student experience at institutional level. In contrast, student experience surveys carried out as part of institutional continuous quality improvement processes, often provide a much fuller and more rounded view of student experiences of higher education at the local level. Over the years, institutional student feedback surveys have used a variety of models but increasingly, they have tended to follow the NSS model in order to allow direct comparisons between the experiences of final-year students and those of students at lower levels (Harvey, 2003). Unfortunately, this has led to a decline in the variation and diversity of the data to be gleaned from such surveys.

Earlier surveys tended to be much more detailed and usually much more tailored to the individual institution. Many student feedback surveys, carried out during the late 1990s and 2000s have used the student satisfaction approach (outlined by Harvey et al., 1997). This approach utilises detailed questionnaires using dimensions that are drawn from discussions with students drawn from a 'group feedback strategy' (Green et al., 1994). The survey is therefore closer than the NSS model to the lived experience of the students in the institution because they are indirectly involved in producing it.

There are many institutional publicly available survey reports that have used this approach. Some have been routinely published, such as those of the University of Central England (UCE) until 2007 and those of Sheffield Hallam University (SHU) in the period 2000–2007. Some others have been published on the Internet, such as those of the University of Greenwich. Most have remained as unpublished reports for internal consumption alone.

Response rates of engineering students

Some of the limited work available on student feedback suggests that engineering students are amongst the poorest at responding to student feedback surveys, both nationally and at institutional level. Nationally, engineering students have been amongst the poorest responders to the NSS. Surridge (2007) observed that there has generally been a lower response rate amongst engineering and technology students than in other subject areas. For example, in 2005, the response rate was 54.5 per cent and in 2006 it dropped to 50.5 per cent. This situation has been reflected at institutional level since the mid-1990s, by which time a number of UK institutions were carrying out regular student feedback surveys. At a north western university in 1998, engineering students provided the lowest response rate, namely 15 per cent in design and technology; 9 per cent in science faculty. At a Welsh university in 2001, the engineering faculty gave a 13.5 per cent response rate, which was lowest equal to that of the school of electronics. At the same institution in 2002, the engineers achieved the lowest response rate per school, namely 12.4 per cent. At SHU (MacLeod-Brudenell et al., 2003), engineering was one of the lowest responders (23.6 per cent), although the lowest was computing and management sciences (21.6 per cent). At UCE, engineering students were consistently the poorest responders to the institutional student satisfaction survey for many years (MacDonald et al., 2005: 3). At another Midlands university in 2005, the engineering response rate was the second lowest in 2005 and lowest of all the faculties in 2007.

There has been no published discussion as to why this may be but there are several possibilities. There may be a simple technological answer: engineering students may be more willing to complete electronic surveys than paper-based ones. However, experience does not always bear this out: at the UCE, an electronic survey replaced the paper version but response rates did not noticeably increase, despite claims that the students would be more willing to complete an electronic survey (MacDonald et al., 2005). The lower response rate may simply be a difference in discipline: better response rates are often achieved by social science students, who are generally more used to completing surveys as part of their studies.

Engineering students in the NSS

The NSS results, which are presented as percentage agreement ratings, are published each year on Unistats,[3] a publicly available website. A fuller analysis of the survey data has been provided by Surridge (2006; 2007; 2008). The full analysis does not need to be repeated here but the few references to engineering students are worthy of mention. The survey, in operation since 2005, finds that engineering students are generally no less positive than other students about the different aspects of their experience. Indeed, only in the teaching and learning section of the survey have engineering students been markedly less positive than other respondents (Surridge, 2006: 77).

The only noticeable differences are related to specific demographic characteristics, and these relate only to specific areas covered by the survey: teaching and learning, assessment and feedback and 'overall satisfaction'. The two main demographic influences noted by Surridge were ethnicity and age. Respondents from Asian, mixed and 'other' backgrounds were less positive about assessment and feedback than others. Asian respondents were less positive about items relating to learning resources; engineering respondents of African descent were more positive than white students about their overall satisfaction (Surridge, 2006: 116, 123). Surridge found that engineering students were also unusual in displaying no age-related effects in their overall satisfaction and in teaching and learning items (Surridge, 2006: 110, 126).

Surridge has also observed an effect of location on ratings from engineering students. Engineering students who were studying off-site were consistently less positive about the organisation and management items of their courses than students in those subject areas whose teaching was provided wholly by their home institution (Surridge, 2006: 119).

In most of these cases, however, it is important to note that students in engineering and technology courses were not unique. Their responses were similar to those in biological science, computer science, business and administration and creative arts and design courses. In some cases, they were more positive than their counterparts who were studying law.

Engineering student feedback in institutional surveys

At institutional level, student satisfaction survey responses suggest that engineering students differ little from students from other disciplines in their satisfaction with aspects of their experience. In some particular areas, engineering students are noticeably less positive than other students but this appears to be more a result of institutional circumstances rather than fundamental discipline-related issues.

Overall satisfaction

Overall satisfaction with the course varies at different institutions: engineering students are slightly less satisfied than others. For example, at UCE in the period 1992–2007 engineering students were consistently less satisfied than other respondents with this aspect of their experience. Engineering students also consistently provided the lowest rating for the university management, for their faculty and for their department. At another Midlands university, ratings from engineering students in 2007 were lower than others for the overall evaluation items: 'university overall', 'value for money' and 'your course' but higher than most for 'the university is enhancing your career prospects'[4] (University B, 2007: 5). At SHU in the period 2000–2003, engineering students were moderately satisfied with their overall experience of the university in comparison with other students (Sheffield Hallam University, 2003: 2).

Similarly, data from the 2009 student satisfaction survey at a university in the south east demonstrates no noticeable difference between engineering students and their counterparts in other disciplines. There is little in the existing data to indicate what is behind engineering students' overall satisfaction but the variety suggests that individual circumstances determine satisfaction in this case.

Interpersonal skills

Engineering students appear to have been slightly less satisfied with the development of interpersonal skills than other students at some universities. At the UCE, for example, satisfaction with this item noticeably declined in the period 1999–2003, when the item was one of the few at the university that gained a C rating[5] (MacDonald et al., 2004: 32, 43). However, satisfaction rose steadily for the next few years and by 2007, the item gained a B rating and was similar to some other faculties (MacDonald et al., 2007: 55, 65). At another Midlands university, most items concerning interpersonal skills were rated as less satisfactory than they were by other students, although engineering students were noticeably more satisfied with the items 'financial management skills' and 'project management skills'. At SHU, interpersonal skills were not regarded by engineering students as any less satisfactory than they were by other students (Sheffield Hallam University, 2004: 23, 41). At a Welsh university, interpersonal skills were not regarded as noticeably different by engineering students compared to students of other disciplines. Again, the variety suggests that local circumstances may determine the satisfaction or otherwise of engineering students.

Work experience and development of employability

Engineering has long had a more direct link to the workplace than many other subjects, thus it may be expected that satisfaction would be linked with aspects of work readiness. Students have expressed concern that they have not been making the links they expected. At a Midlands university, opportunities to make contact with professionals was considered to be of lower importance by engineering students than by those of other faculties. Work experience items were all regarded lower than by other students. At a Welsh university, the organisation of work experience was regarded as unsatisfactory and very important (D rating) in 2001, and engineering was amongst the least satisfied of all faculties. However, the following year, this item was regarded as satisfactory and very important (B rating). At SHU in 2003, the item 'opportunities for work-related placements' was regarded as satisfactory and very important (B rating) by engineering students, as it was by most students of other disciplines. However, the separate item 'opportunities for activities related to enterprise and/or self-employment' was regarded as adequate and not so important (Sheffield Hallam University, 2003: 23). The following year, items relating to work experience were not regarded as any less satisfactory by engineering students. At UCE, the two items 'suitability' and 'organisation of work experience' suffered a collapse in satisfaction in 1999 and only a slow recovery until 2004 (MacDonald et al., 2004: 42). By 2007, 'suitability' and 'organisation of work experience' were both regarded as very satisfactory or satisfactory, respectively, and very important (ratings A and B) (MacDonald et al., 2007: 55).

This was clearly an issue for the Faculty of Engineering and it made efforts to address the concerns raised by students in the 2000 report, when the item 'opportunities to make links with professionals' was regarded as unsatisfactory and very important (D rating) by engineering students:

> The Faculty of Engineering and Computer Technology will continue to advertise trips organised through the Engineering Society and will continue to support external initiatives for the benefit of students. Students will continue to be provided with *opportunities to link with employers* through evening events and other schemes. (Bowes et al., 2000: 12)

'Work experience' items continued to do less well in engineering than in other faculties at UCE, but, in 2000, the opportunities to make links with professionals was regarded as satisfactory and very important (C rating) and the faculty continued to address the concern:

> The Faculty of Engineering and Computer Technology will continue to provide publicised, evening events at which students may meet employers, and will encourage more students to attend. There are also opportunities to link with industry through the CISCO academy.[6] (Harvey, 2001: 10)

The faculty continued to address this issue in 2004 and 2005, investigating potential opportunities for facilitating work experience, where possible or appropriate, working with central offices to provide a more effective approach:

> The Faculty of Engineering [Technology Innovation Centre] (tic) is working with the Careers Services to improve placement preparation and placing more opportunities on the *tic* intranet. Events are planned in the tic café to advertise the career services. (MacDonald et al., 2004: 10)

> The support for students who wish to undertake a work placement has been improved by increasing the staffing of the placements office, providing more placement preparation seminars and activities and publishing more placement opportunities on the tic's placement website. (MacDonald et al., 2005: 12)

Work experience is arguably an area that is less likely to be regarded as satisfactory and as a result, harder to improve.

Assessment and feedback

The area where engineering and technology students appear to have been least satisfied over the years is that of assessment and feedback and this reflects the national concern with this area (Williams and Kane, 2008). The usefulness of feedback usually scores more highly than the promptness of feedback but in both, science-related disciplines and especially engineering and technology generally score less well for both items than others. It should also be noted that few faculties score particularly well. This indicates that assessment and feedback may be particularly difficult issues to improve in engineering subjects.

The reasons for such intransigency are not, however, clear from existing evidence. Indeed, as Elton noted (1998: 36), this may be related to structural differences in the form of assessment in 'hard' subjects. Indeed, from the qualitative data collected from student experience surveys, engineering students reflect concerns that are common to students in many other disciplines. More specifically, students' comments suggest that the tutors' responses to their assignments give them an indication of their progress. Many such comments are not principally about feedback but about the marks themselves; although it is not always clear if students' use of the word 'mark' refers solely to the grade or to the comments and grade applied to the assignment by the tutor as the following quotes indicate:

> We need our marks to indicate our progress. By not having any, we have no way of knowing how well/badly we are performing![7]

> We need to know how well we did.[8]

> Coursework should be published on the Internet. What's the point of having a 'mark' column when the marks are not published on the Internet? All coursework marks should be put up there.[9]

This is despite Gibbs' (2006a: 34) view that:

> . . . if students receive feedback without marks or grades, they are more likely to read the feedback as the only indication they have of how they are progressing.

Some comments make it clear that it is marks the students are concerned about – real concern with the most basic unit. Other comments demonstrate a deeper understanding of the issue – that a mark is not necessarily self-explanatory:

> Very difficult to understand what they say . . . . [H]ow will I improve my assignments. I don't know why I fared bad[ly] in my test!![10]

> Feedback on coursework: how can a lecturer justify a relatively low mark with simply a collection of ticks? A complete lack of respect has been shown by a number of lecturers generally.[11]

Promptness of feedback is an issue because many students do not feel that feedback is returned quickly enough. In many cases, there is evidence of irritation amongst students that they are expected to hand in their work on time but that staff do not return work on time:

> Why can work be returned late, but you are not allowed to hand it in late?[12]

> Several times this year I have had very late feedback for coursework that was handed in on time.[13]

> Lecturers need to mark and return work on time! After all, we get penalised for late submission, so what happens if they don't get it done?[14]

Assessment and feedback items are consistently regarded as very important by respondents, although closer analysis of UCE data demonstrates that they were in fact regarded as consistently less important by engineering students than others. Art school students ranked the usefulness of feedback particularly highly in 1996[15] and acting (drama) students regarded it particularly highly in 2007.[16] For business school students, the importance of this issue has increased considerably, although the number of items has decreased. Engineering students have not regarded this item as important.[17] Education students regarded the promptness of feedback as most important in 1996[18] and highly in 2007.[19] Health students regarded it as most important in 2007.[20] Engineering students regarded it as least important in both 1996 and 2007[21] (Williams and Kane, 2008).

Institutional responses

A fundamental element of the student satisfaction approach is that action based upon the survey results is planned and implemented (Harvey et al., 1997, Part 10). Assessment and feedback is often mentioned in institutional feedback to staff and students as an area in which action is being taken. At UCE:

> ... the issue of promptness of feedback on assignments has always been very important to students and satisfaction varies from faculty to faculty and from course to course. (Harvey et al., 2000: 11)

Publicly available feedback information from UCE demonstrates that the issue of feedback as it is raised by students has been addressed by management in several different ways since the mid-1990s. Faculties that have scored badly on assessment issues have responded by setting realistic targets for assignment turn-around and, importantly, made assessment feedback schedules clearer to their students. Communication is a consistent feature of action. This is combined with closer monitoring of actual turnaround times.

Institutional action plans seldom refer directly to innovative practice as it is described in the literature (e.g. Stefani, 2004–2005; Mutch, 2003). Approaches to assessment and feedback are used that are recognisable as part of a wider trend, such as computer-assisted learning.

Responses to institutional surveys mainly fall into two categories:

- The institution clarifies its procedures to the students.

- The institution recognises that it can improve its own processes.

The most important element in an institution's response to student feedback is that transparent action is taken. Although a causal link is difficult to prove, it is clear that a rise in satisfaction with an item often coincides with action as a result of the annual survey. In other words, the result of using student data in order to inform improvements appears to have a direct impact on the resulting student satisfaction. Satisfaction is therefore a dynamic process that depends on institutions asking for feedback from their students and acting upon the information. Furthermore, students need to be made aware of the action that has been made so that they can see that the feedback process is worthwhile and not merely an empty gesture.

One response to 'promptness of feedback on assignments' is to set realistic targets. In 1995–1996, for example, UCE instituted more realistic turn-around targets:

> In response [to low satisfaction with promptness of feedback] the Faculty [of Computing and Information Studies] set a target of a 'four-working-week turnaround' on assignments, which has proved very successful. (Centre for Research into Quality, 1996)

At SHU, there is a three-week rule for the return of feedback to students (Harvey et al., 2004), although it has been made clear, publicly, that this is difficult to achieve in some faculties.

In addition to setting realistic targets for the return of feedback on assignments, an important element in the institutional response is to make the schedule of feedback on assignments clearer to students. At the UCE in 2002–2003, there was an attempt to clarify hand-in dates:

> In order to enable students to plan their workload, the Faculty of the Built Environment is to identify assessment dates clearly and no longer allow alterations of hand-in dates. (MacDonald et al., 2003: 11)

At UCE over the period 1995 to the early 2000s, there was a strategy to provide students with clear guidelines on what to expect. This was aimed not only at informing students about deadlines but to enable them to plan their workload. Indeed, in 1995–1996, faculties such as Health, Social Science, Engineering and Computer Technology wrote assignment turn-around guidelines into their student charters. All this relates to communication with students about what is expected of them.

Better communication on all aspects of assessment is an important issue for students, as argued by Price and O'Donovan (2006), and this is taken into account by institutions that manage a feedback and action cycle. At SHU in 2005, the Faculties of Arts, Computing, Engineering and Sciences offered guidance to staff about providing feedback to students on the virtual learning site and published it in the faculty newsletter (SHU, 2005).

In addition to communication and the clarification of deadlines, better timetabling of assessments is thought by some institutions to help relieve the problems of promptness of feedback.

> The tic has introduced a new system to aid the tracking of coursework and ensure that work is returned according to the tic's published timescales. In future, students will be notified by email when work is ready for collection from the Learning Centre. This will hopefully address the current situation where work has been available for collection but students have not been informed. (MacDonald et al., 2005: 11)

Ensuring an even spread of assessments is one approach that has been used by institutions. At UCE over several years, attempts were made to spread assignments more evenly by changing the teaching programme. At another Midlands university in 2005–2007, attempts have been made to avoid 'bunching' of assessments.

At UCE's Faculty of the Built Environment in 2003–2004, a new postgraduate framework offered a less frequent assessment régime. At another Midlands university, reports from external examiners stated that the university is assessing too much. The response has been to change courses from a 15- to a 20-credit framework. This is reflected in the wider discussion about assessment: modular structure has led to shorter courses and, as a result, more frequent assessment (Gibbs, 2006b).

At this Midlands university, the internal audit indicated that there are two basic pedagogies about assessment:

- assessment tests what students have learnt;
- assessment forces students to learn: by reading a lot.

This reflects the observation by Knight (1995) that assessment is an effective method of making students work.

> Overall, the issue of promptness of feedback has been a dominant one for students. Faculties generally explain that this might be because marks are no longer given out until moderated, as changes in assessment marks sometimes take place. Faculties have acted where possible and are investigating the issue. The tic is going to pilot a new system aimed at accelerating the return of marked work

in order to enable students to plan their workload. (MacDonald et al., 2003)

A common response to assessment and feedback issues is to institute an effective monitoring system. Different faculties at UCE used different approaches. For example, in the Faculty of Education, in 2005–2006 monitoring of the timing and placing of assessments was carried out in order to make improvements in this area. In 1998–1999 in the Faculty of Engineering, selected modules were audited and students were asked specifically to comment on this issue. The Board of Studies was charged with determining those modules where promptness of feedback was a problem.

In addition, some faculties institute systems to track student coursework to ensure that feedback is provided according to schedules.

> The management information systems have caused some delays for students and these are currently under review. (Bowes et al., 2000: 12)

> At the Technology Innovation Centre (tic), improvements have been made to the coursework hand-back system and on-line tracking of coursework is now available to students. (MacDonald et al., 2006: 12)

Institutions have tried to increase the immediacy of providing feedback to students. For example, at the UCE Faculty of Engineering and Computer Technology in 1999–2000, management emphasised the use of assessment in class so that students receive immediate feedback. In the School of Property and Construction in the Faculty of the Built Environment that year, there was a commitment to provide general feedback to students within two weeks, with the individual pieces of course work being returned within a further two weeks.

> More emphasis is now being given to assessment in class so that students receive immediate feedback. (Bowes et al., 2000: 12)

In some cases, the problem has been simply a lack of academic staff. For example, at UCE, several faculties increased the number of teaching staff and found that promptness of feedback was less of a concern to students. It is generally recognised that such a simple response is not easy in the context of a hugely expanded higher education sector where many of the traditional one-to-one methods of teaching have proved too resource intensive (Gibbs, 2006b: 12). Nonetheless, there is a case for assigning more of an academic's time to feedback given its importance in the learning process.

> In addition, the Faculty has employed more visiting tutors to cope with staff shortages and is undertaking a pilot to provide direct feedback using e-mail. (Bowes et al., 2000: 12)

Increasingly, institutions are exploring the use of alternative methods of feedback to students such as those explored by Bryan and Clegg (2006). Of particular interest currently is the potential for computer-assisted assessment (Swain, 2008). This is in part a response to the Higher Education Funding Council for England's 2005 strategy for e-learning, but many institutions have been developing electronic methods of assessment and feedback for many years. At UCE, for example, the use of electronic modes of feedback has been developing for some years. In 1999–2000, the Faculty of Engineering and Computer Technology undertook a pilot to provide direct feedback using e-mail. Pilot tutorials using the Internet provided more immediate feedback. By 2003–2004, the tic intranet provided information on extenuating circumstances, appeals procedures and degree classification. Students could now obtain module marks online and download details on any work required for re-assessment. In 2004–2005, staff in the Faculty of the Built Environment were given further training on the university's electronic information system. In the Faculty of Engineering:

> Learning support will be provided through learning materials on the Internet. (Bowes et al., 2000: 12)

> Pilot tutorials using the Internet will provide more immediate feedback. (Bowes et al., 2000: 12)

In some cases, institutions attempt to involve students more openly in the feedback return process. At UCE in 2000–2001, students were employed by one faculty to act as co-ordinators. This approach, it was believed, would help to alleviate difficulties in receiving work from visiting tutors, more of whom have been recruited. The faculty reported improvements to the system and students are now able to retrieve their end-of-semester results through the intranet. This indicates an understanding that the students themselves can be a useful resource in the assessment process (Falchikov, 1995).

A frequent element in institutional approaches is to introduce standardised feedback systems. At UCE, for example, in 2001–2002 the

Faculty of Law and Social Sciences proposed a new form to facilitate feedback relating to intended learning outcomes and to identify areas in which students can improve academic performance. In 2006–2007, the UCE Business School reviewed the timing of assessments and the overall assessment strategy (for first-year students) in order to support students through the initial challenges of entering higher education. It also reviewed its systems for handing in and returning assignments and set up an assigned 'hand-in/collection' point specifically for this purpose.

Satisfaction with aspects of assessment and feedback are sometimes attributed to what were regarded as external factors. At UCE, for example, faculties generally explained poor satisfaction with 'promptness of feedback on assignments' as being due to marks being no longer given out until moderated, as the assessment marks are sometimes changed by this process.

Some institutions attempt to draw together examples of good practice internally. At University A, for example, an audit committee set up in 2004 to establish what approaches were being used by high-scoring faculties within the university could be adopted institution-wide. The audit recommended that:

> . . . a more formalised policy governing the provision of feedback to students on examinations should be developed and that action be taken to ensure that both staff and students are made fully aware of the faculties' expectations in this area.
> . . . mechanisms already established for the provision of feedback on examinations should continue to be developed and that they be clearly communicated to staff and students.
> . . . mechanisms to improve the timescales for the return of feedback on assessed work should continue to be investigated and developed and that measures be taken to ensure that the communication channels for providing information to students in respect of these timescales work as effectively as possible.
> University guidance should be provided for faculties governing the electronic provision of feedback on draft work.

A similar approach was taken at University B as a result of its 2005 survey:

> A new policy on coursework submission, return and feedback has been put in place, reflecting the good practice exhibited in many areas and will be implemented fully for the 2007–2008 academic year.[22]

Students, not only those from engineering and technology faculties, do not always collect work when it is marked and available on time. At UCE, for example, one faculty dean observed that whilst it is relatively easy to implement and enforce collection points for assignments to be handed in, it is difficult to enforce the collection of marked work by students. His experience had been that when return points had been designated, many assignments remained uncollected. This experience was also reflected at SHU in 2002:

> Pilot work-return sessions were organised by the administration team but many students did not use the opportunity to collect work. The pilot scheme will be repeated in 2002–2003. (Morey et al., 2002)

Similarly, at University B, up to 30 per cent of coursework remained uncollected, in part the pro-vice chancellor observed, because marks were available online.

This raises two questions. First, students may be indifferent to comments and only want grades although this does not seem reasonable in the light of comments from many students wanting feedback so as to improve their next assignment. It is arguable that increasing use of modules and semesters has created a situation where there is no further possibility to improve as there is often one summative assessment in a subject so students do not want the feedback as (they perceive that) it is non-transferable (Gibbs, 2006b: 11).

Second, it may be the case that the coursework is not collected because it is not seen as valuable. Collection of feedback may not be viewed by the students as a beneficial process because it means that they read the comments without the dialogue necessary to explain, explore or contest the comments. Worse, students may not collect feedback because they believe that the lecturers' comments are insignificant, being scanty and adding nothing to the grade.

Issues relating to the multi-campus university

One of the major issues affecting student satisfaction amongst engineering students at UCE appeared to have been a significant change in location and it is useful to explore this particular example in detail. In 2001, the Faculty of Engineering moved from the main campus at Perry Barr to Millenniun Point, a new prime site in the city centre.[23] Although this was heralded as a great new opportunity, satisfaction amongst the university's engineering students took a tumble in almost all the dimensions covered by the student satisfaction survey for the next few years.

One of the issues that clearly emerges from the surveys is distance created between the new campus and the main campus. Both the Students' Union facilities and Student Services received particularly low ratings, presumably because they were seen as remote from the new campus. In particular, students complained about the lack of information about Student Services and the lack of visibility of these services on the new campus. This has been a perennial problem since the faculty moved to the new site.

There was a concerted effort by UCE in 2004 to improve the experience of engineering students at the city site by introducing several new measures. First, the Student Union:

> . . . improved transport arrangements this year from campuses in order to improve access for tic and other, smaller campuses. (MacDonald et al., 2004)

Second, to make the union space at the faculty more visible:

> . . . the layout and facilities in the Union Student room at tic has been revised, including 'relaxation' facilities such as widescreen TVs and easy chairs. The Union of Students holds lunchtime events in the tic café. (MacDonald et al., 2004)

> In 2004, to raise awareness about the availability and diversity of Student Services; Services have also increased the number of hours of support at tic. (MacDonald et al., 2004).

Did these efforts work? It is difficult to prove causal links but there appears to have been some improvement at UCE over the course of the following few years. In 2005 and 2007, respondents from the new city site were no less positive about aspects of the Student Union and social life than other students generally (MacDonald et al., 2005, 2007).

Conclusions

Engineering departments have suffered from poor recruitment and poor retention rates in the UK for many years but an exploration of student feedback surveys has not clearly identified a particular area of concern in the lived experience of engineering students. Rather, the overall experience of engineering students appears to be little different from that of their fellow students in other faculties. Indeed, they seem generally to be moderately satisfied with their experience, especially when compared with the experiences of, for example, art and design students (Yorke, 2008).

There are only two noticeable and consistent differences between engineers and their fellow students. First, they are less responsive to feedback surveys. This may be the result of the discipline itself. Engineering students are not expected to use questionnaires as part of their training whereas, for example, social science students, who are generally much better at responding, often work with questionnaires. Second, engineering students appear to rate assessment and feedback worse compared to students of other disciplines, suggesting that there are structural issues with the type of assessment usually performed in these subjects (Elton, 1998).

However, in some institutions, there are specific areas in which engineering students are less satisfied than others, although these issues do not coincide at different institutions. Indeed, poor satisfaction levels appear to arise as a result of local circumstances. In many cases, this may be the result of policy decisions at the institutional level and are unlikely to be connected with the discipline in particular. It is noticeable that when concerted action is taken by an institution as a result of listening to the student voice – and such action is prolonged and well communicated – then satisfaction tends to increase over time.

This has important implications for our understanding of the role of student feedback as part of a quality improvement process. Where student satisfaction surveys have been conducted consistently for several years, it is possible to see changes in satisfaction with different aspects of the student experience (Williams and Kane, 2009). Not only can downward trends be clearly identified, but upward trends coincide with action on the part of the institution taken as a result of listening to the students.

The key issue in collecting student feedback, therefore, is what is done with it. Student feedback surveys are not merely measurement tools but are dynamic instruments that need to be used in combination with an institutional quality improvement process. Understanding local circumstances is vital in improving local situations.

Notes

1. A social research unit based at the then University of Central England, now Birmingham City University.
2. The NSS was introduced in the UK in 2005 as a measurement of the quality of higher education from the students' perspective. The questionnaire,

delivered electronically, contains a relatively small number of dimensions (usually 20–30 questions) and uses a Likert scale of agree/disagree. The survey has, since its inception, been highly controversial yet extremely influential on the policies and strategies of higher education institutions (see Williams and Cappuccini-Ansfield, 2007).

3. Available online at: *http://unistats.direct.gov.uk/* (accessed October 2011).
4. To protect confidentiality, two universities have been anonymised as 'University A' and 'University B'. The report cited here comes from University B (2007: 5) and has not been published.
5. The student satisfaction approach, developed by Harvey et al. (1997), is unusual in combining satisfaction and importance ratings as an alphabetical score. In this scheme, A = very satisfactory, B = satisfactory, C = adequate, D = unsatisfactory and E = very unsatisfactory; the letter case represents levels of importance to the students. Hence, capitalised letters indicate that respondents regard an item as very important, lower case letters that an item is important and lower case scores in parenthesis are not so important.
6. An education initiative from the US global conglomerate, Cisco Systems, offering a range of IT networking programmes aimed at preparing students for the eponymous certification exams.
7. Student at University B, Faculty of Engineering, 2007.
8. Student at University A, Faculty of the Built Environment, 2005.
9. Student at University A, Faculty of Engineering, 2007.
10. Student at University A, Faculty of Engineering, 2007.
11. Student at University A, Faculty of Law and Social Science, 2007.
12. Student at University A, Faculty of the Built Environment, 2000.
13. Student at University B, Faculty of Business, 2005.
14. Student at University B, Faculty of Engineering, 2005.
15. Mean importance = 6.51.
16. Mean importance = 6.64.
17. Mean importance = 6.07 in 1996 and 6.08 in 2007.
18. Mean importance = 6.32.
19. Mean importance = 6.25.
20. Mean importance = 6.39.
21. Mean importance = 5.79 in 1996; mean importance = 5.89 in 2007.
22. University B, 2005.
23. Millennium Point is a large multi-purpose building in a regeneration zone in the heart of Birmingham. It was built as a major millennium project in 2000 and houses the city's science museum, ThinkTank, and the UCE Engineering Faculty.

References

Bowes, L., Harvey, L., Marlow-Hayne, N., Moon, S. and Plimmer, L. (2000) *The 2000 Report on The Student Experience at UCE*. Birmingham, UK: University of Central England.

Bryan, C. and Clegg, K. (eds.) (2006) *Innovative Assessment in Higher Education*. London: Routledge.

Centre for Research into Quality (1996) *Student Satisfaction*. February 1996. Birmingham, UK: University of Central England.

Confederation of British Industry (2008) *Taking Stock CBI Education and Skills Survey 2008*. London: Confederation of British Industry and edexcel.

Elton, L. (1998) Are UK degree standards going up, down or sideways? *Studies in Higher Education*, 23(1), 35–42.

Falchikov, N. (1995) 'Improving feedback to and from students'. In P. Knight (ed.) *Assessment for Learning in Higher Education*, pp. 157–66. London: Kogan Page.

Felder, R.M., Felder, G.N. and Dietz, E.J. (2002) 'The effects of personality type on engineering student performance and attitudes'. *Journal of Engineering Education*, 91(1), 3–17.

Flynn, S. (2008) 'Engineering students suffer highest drop-out rates, say MPs'. *New Civil Engineer*, 20 February 2008. Available online at: *http://www.nce. co.uk/engineering-students-suffer-highest-drop-out-rates-say-mps/756664. article* (accessed October 2010).

Gibbs, P. (2006a) 'How assessment frames student learning'. In C. Bryan and K. Clegg (eds.), *Innovative Assessment in Higher Education*, pp. 23–36. London: Routledge.

Gibbs, P. (2006b) 'Why assessment is changing'. In C. Bryan and K. Clegg (eds.), *Innovative Assessment in Higher Education*, pp. 11–22. London: Routledge.

Green, D. with Brannigan, C., Mazelan, P. and Giles, L. (1994) 'Measuring student satisfaction: a method of improving the quality of the students' experience?' In S. Haselgrove (ed.), *The Student Experience*. Buckingham, UK: Society for Research into Higher Education and Open University Press.

Gregory, R., Thorley, L. and Harland, G. (1994) 'Using a standard student experience questionnaire with engineering students: initial results'. In G. Gibbs (ed.) *Improving Student Learning: Theory and Practice*. Oxford, UK: Oxford Centre for Staff Development.

Harvey, L. (2001) *The 2001 Report on the Student Experience at UCE*. Birmingham, UK: University of Central England.

Harvey, L. (2003) Student feedback. *Quality in Higher Education*, 9(1), 3–20.

Harvey, L., Ibbotson, R., Leman, J. and Marsden, D. (2004) *Sheffield Hallam University Student Experience Survey 2004: Undergraduate and Taught Postgraduate Students*. Sheffield, UK: Sheffield Hallam University.

Harvey, L., Moon, S. and Plimmer, L. (1997) *Student Satisfaction Manual*. Buckingham, UK: Society for Research into Higher Education and Open University Press.

Higher Education Funding Council for England (HEFCE) (2002) *Information on Quality and Standards in Higher Education*. London: HEFCE. Available online at: *http://www.hefce.ac.uk/pubs/hefce/2002/02_15/02_15.pdf* (accessed November 2010).

Higher Education Funding Council for England (2005) *HEFCE strategy for e-learning*. March 2005/12. Available online at: *http://www.hefce.ac.uk/pubs/ hefce/2005/05%5F12/05_12.pdf* (accessed October 2011).

HM Treasury (2002) *Investing in Innovation: A Strategy for Science, Engineering and Technology*. London, HM Treasury. Available online at: *http://webarchive. nationalarchives.gov.uk/+/http://www.hm-treasury.gov.uk/media/F/D/science_ strat02_ch1to4.pdf* (accessed October 2010).

Knight, P.T. (ed.) (1995), *Assessment for Learning in Higher Education*. London: Kogan Page.

MacDonald, M., Saldaña, A. and Williams, J. (2003) *The 2003 Report on the Student Experience at UCE*. Birmingham, UK: University of Central England.

MacDonald, M., Schwarz, J., Cappuccini, G., Kane, D., Gorman, P., Sagu, S. and Williams, J. (2005) *The 2005 Report on the Student Experience at UCE*. Birmingham, UK: University of Central England.

MacDonald, M., Williams, J. and Saldaña, A. (2004) *The 2004 Report on the Student Experience at UCE*. Birmingham, UK: University of Central England.

MacDonald, M., Williams, J., Gorman, P., Cappuccini-Ansfield, G., Kane, D., Schwarz, J. and Sagu, S. (2006) *The 2006 Report on the Student Experience at UCE*. Birmingham, UK: University of Central England.

MacDonald, M., Williams, J., Kane, D., Gorman, P., Smith, E., Sagu, S. and Cappuccini-Ansfield, G. (2007) *The 2007 Report on the Student Experience at UCE*. Birmingham, UK: University of Central England.

MacDonald, M., Williams, J., Kane, D., Gorman, P., Smith, E., Sagu, S. and Cappuccini-Ansfield, G. (2007) *The 2007 Report on the Student Experience at UCE*. Birmingham, UK: University of Central England.

MacLeod-Brudenell, T., Ibbotson, R., Smith, M., Harrison, A., Harvey, L., Leman, J. and Fowler, G. (2003) *The 2003 Report on the Undergraduate and Taught Postgraduate Student Experience at Sheffield Hallam University*. Sheffield, UK: Sheffield Hallam University.

Morey, A., Watson, S., Saldaña, A. and Williams, J. (2002) *The 2002 Report on the Undergraduate and Taught Postgraduate Student Experience at Sheffield Hallam University*. Sheffield, UK: Sheffield Hallam University.

Mustoe, L.R. and Croft, A.C. (1999) 'Motivating engineering students by using modern case studies'. *International Journal of Engineering Education*, 15(6), 469–76.

Mutch, A. (2003) 'Exploring the practice of feedback to students'. *Active Learning in Higher Education*, 4(1), 24–38.

Parsons, S.J. (2003) 'Overcoming poor failure rates in mathematics for engineering students: a support perspective'. Available online at: *http://www.hull.ac.uk/engprogress/Prog3Papers/Progress3%20Sarah%20Parsons.pdf* (accessed October 2010).

Price, M. and O'Donovan, B. (2006) 'Improving performance through enhancing student understanding criteria and feedback'. In C. Bryan and K. Clegg (eds.) *Innovative Assessment in Higher Education*. London: Routledge.

Richardson, J.T.E. (2005) 'Instruments for obtaining student feedback: a review of the literature'. *Assessment and Evaluation in Higher Education*, 30(4), 387–415.

Sazhin, S.S. (1998) 'Teaching mathematics to engineering students'. *International Journal of Engineering Education*, 14(2), 145–52.

Sheffield Hallam University (2003) *The 2003 Report on the Student Experience at SHU*. Sheffield, UK: Sheffield Hallam University.

Sheffield Hallam University (2004) *The 2004 Report on the Student Experience at SHU*. Sheffield, UK: Sheffield Hallam University.

Sheffield Hallam University (2005) *The 2005 Report on the Student Experience at SHU*. Sheffield, UK: Sheffield Hallam University.

Souitaris, V., Zerbinati, S. and Al-Laham, A. (2007) 'Do entrepreneurship programmes raise entrepreneurial intention of science and engineering students? The effect of learning, inspiration and resources'. *Journal of Business Venturing*, 22(4), 566–91.

Stefani, L. (2004–2005) 'Assessment of student learning: promoting a scholarly approach'. *Learning and Teaching in Higher Education*, 1(1), 51–66.

Surridge, P. (2008) *The National Student Survey 2005–2007: Findings and Trends*. Bristol, UK: Higher Education Funding Council for England.

Surridge, P. (2007) *The National Student Survey 2006 Report to HEFCE*. Bristol, UK: Higher Education Funding Council for England.

Surridge, P. (2006) *The National Student Survey 2005 Report to HEFCE*. Bristol, UK: Higher Education Funding Council for England.

Swain, H. (2008). 'Evaluating students via online assessment both tests what students know and helps develop their understanding'. *Times Higher Education Supplement*, 03 January 2008. Available online at: *http://www.timeshighereducation.co.uk/story.asp?sectioncode=26&storycode=21005 1&c=1* (accessed October 2011).

Symons, R. (2006) 'Listening to the student voice at the University of Sydney: closing the loop in the quality enhancement and improvement cycle'. Paper presented at the *2006 Australian Association for Institutional Research (AAIR) Forum*, Coffs Harbour, New South Wales, November 2006.

Williams, J. and Cappuccini-Ansfield, G. (2007) 'Fitness for purpose? National and institutional approaches to publicising the student voice'. *Quality in Higher Education*, 13(2), 159–72.

Williams, J. and. Kane, D. (2008) *Exploring the NSS: Assessment and Feedback Issues*. York, UK: Higher Education Academy.

Williams, J. and Kane, D. (2009) 'Assessment and feedback: institutional experiences of student feedback, 1996 to 2007'. *Higher Education Quarterly*, 63(3), 264–86.

Williams, R. and Brennan, J. (2003) *Collecting and Using Student Feedback on Quality and Standards of Learning and Teaching in Higher Education*. Bristol, UK: Higher Education Funding Council for England.

Yorke, M, (2008) 'What can art and design learn from surveys of "the student experience"?' Paper presented at the *Conference of the Group for Learning in Art and Design*, Nottingham, UK, 8 September 2008.

8

Trends, issues and the future of student feedback in engineering

Chenicheri Sid Nair, Patricie Mertova and Arun Patil

Abstract: This chapter draws on the earlier chapters in this book about student feedback in engineering education by several international contributors. It summarises the common themes, trends and issues within the discipline with respect to student feedback. It concludes with some thoughts on the future directions of student feedback within the discipline.

Key words: student feedback, engineering education, enhancement of learning and teaching.

Introduction

Research over the past four decades has consistently shown the importance and relevance of student feedback in educational settings (e.g. Doyle, 1975; Feldman, 1976; Marsh et al., 1979; Marsh, 1984; Fraser, 1991, 1994; Harvey, 2001; Nair, 2011). Despite the general acceptance that student feedback is important, there has been, and still is, much discussion about what needs to be collected and the use that should be made of the information, once it has been collected. The arguments that are consistently raised concern topics such as:

- the purposes for which the feedback has been collected;
- the types of questionnaires that are being utilised to elicit such feedback;
- the purposes for which the data is to be used once it has been collected;
- the availability of the data to different audiences.

Even fundamentals, such as the validity of student feedback are a recurring theme in many of these discussions.

Despite the continuing debate, there is a growing body of research which argues that student feedback informs universities of issues and student concerns in many areas of university life; and that universities need to engage in ongoing communication with their students (Harvey, 2001; Coates, 2006). As early as 1982, Astin succinctly commented that students, in particular, are in a good position to comment on the courses of study, which in turn will assist institutions not only to improve their teaching and learning but also to contribute to greater personal development for the students (Astin, 1982).

However, student feedback was not commonly used in many institutions until around the mid-1990s. The massification of higher education globally (including the emergence of a number of private providers), and pressures over political and financial control of higher education have lead to a global push to ensure quality in higher education (Green, 1994; Harvey, 1998; Brown, 2004). Following these trends, students gradually started being regarded as one of the primary 'stakeholders' in higher education and thus their voice started being regarded as a significant factor in enhancing the quality of teaching and learning and other aspects of student experience (Harvey, 2001; Richardson, 2005; Coates, 2006; Williams and Cappuccini-Ansfield, 2007). Bennett and Nair (2010) identified a number of drivers for the increased collection of student feedback in higher education. These included:

- diagnostic feedback to aid the development and improvement of teaching;
- research data to inform changes in units, courses, curricula and teaching;
- administrative decision making in terms of teaching and learning;
- information for current and potential students;
- measurement of the quality of courses (in some countries this is tied to further funding).

With importance being placed on such feedback, a variety of tools are currently being employed to collect it. These include: feedback on teaching, unit or course design, course structure and satisfaction with the services provided by the university. Feedback on higher education is also sought from other stakeholders, such as: employers, professional bodies and others (Harvey, 2001; Coates, 2006). Such feedback is gaining prominence. This applies broadly to higher education, including the engineering disciplines which are the focus of this book.

Issues, trends and approaches

Generally, the international perspective on student feedback in engineering education is that it should be part of the quality approach within these engineering disciplines to help improve engineering education. The main argument of this book assumes that such feedback is essential to improve the key learning outcomes of engineering education, such as enabling the skills to understand, communicate and solve problems. This use of student feedback to achieve such an outcome is not new, as general education research or 'knowledge of the context and theories on student learning' (Chapter 1, p. 14) has been utilised to enhance engineering education for some time. It is increasingly recognised that there is a need to understand not only the teaching and learning, but also the total learning environment so as to assist students to develop these skills and attributes in the course of their studies so that they can successfully graduate as engineers.

The emphasis on the necessity for engineering graduates to have acquired soft skills is discussed in detail in a range of national contexts: Sweden (Chapter 1), Chile (Chapter 2), Hong Kong (Chapter 3), India (Chapter 4) and Thailand (Chapter 5). Utilising student feedback to ensure the match of the requirement for soft skills from industrial practice and employers has also been highlighted in a number of earlier studies (e.g. Nair et al., 2009; Patil, 2005; Radcliffe, 2005).

An important subject, which resonates across the majority of the chapters in this book, is the political agendas of different countries to improve the quality of their engineering graduates. This is in line with the quality agenda of improving education in general. However, the active role of professional bodies in ensuring the quality of graduates for a global market is specific to engineering. Such pressures have been critical in institutions putting appropriate student feedback collection mechanisms in place, predominantly under the guise of improving the student experience. Student experience has also been an important driver for collecting feedback by tertiary institutions, however, perhaps not the most dominant one, in the case of engineering.

Improvements resulting from feedback are a recurring theme in a number of the chapters of this book. This sentiment was strongly emphasised in the first book in the series to which this volume belongs, which looked at student feedback in higher education generally (Nair and Mertova, 2011). A common thread that binds the discussion is that student feedback should not be considered to simply be a data generating exercise. But the data has to be utilised for genuine improvement. This argument was eloquently summarised by Kogan (1990):

> In principle, evaluation should not be made at all unless those making or requiring the evaluation are sure how they are going to benefit from it.

Kember et al. (2002) and Edström (2008) further developed this point by arguing that improvement can only achieved if there is a purpose behind the collection of the data, there also needs to be appropriate analysis and interpretation with actions to follow. These authors argue that to achieve all these, there needs to be appropriate resourcing by institutions, including support and leadership.

Apart from the need to have an action plan to improve the situation, a number of chapters in this book discuss the nature of the questionnaires that should be utilised to obtain particular feedback. The value of qualitative in-depth forms of evaluation, which provide a richer insight that is not necessarily given by quantitative approaches, has been strongly argued in several chapters (Chapters 1, 3, 6 and 7). However, the requirement or utility of summative feedback particularly for quality assurance purposes, and the difficulties in using formative feedback have also been recognised (Chapters 1, 3 and 6). From the perspective of enhancing teaching and learning, formative evaluation was highlighted as being more beneficial where feedback is regarded as a continuous process where the teachers are able to use the information to improve their performance, and this in turn benefits the students during their course (Chapters 1, 3 and 6). This point was supported by the argument that educators are more reluctant to recognise the importance of summative evaluations as these are designed to meet administrative and public obligations, instead of as a constructive means of providing continuous, up-to-date, practical benefits to both students and teachers (see Chapter 1 (pp. 1–23) and Chapter 3, pp. 43–59). Related to this, was an argument that the actual experience of the teachers should be included and complement the summative process, such input would add much richer insights to the evaluation of the teaching (Chapter 1, pp. 1–23).

There is also some discussion about the tension between evaluation of the teachers and student learning. Arguments pivot around the paradigms of evaluation of learning versus the evaluation of teaching. The point that has been stressed is that student learning should be the primary concern and if this is so then such evaluation should take precedence in institutions (Chapter 1).

Conclusions

This book provides a rich insight into international perspectives on the hows and whys of student feedback in engineering. The chapters point to recent developments, which have taken place within the discipline, where theories and research on teaching and learning are forming the groundwork necessary to effectively implement and understand student feedback. However, the international perspectives indicate that the implementation of an effective feedback system within the discipline is in its infancy in many parts of the world (see Chapters 4 and 5). Clearly, there is recognition that such feedback will help enhance the quality of the engineering graduates and this is pivotal for any effective change to take place.

The intention of this book was not to provide any 'recipes' for achieving optimal outcomes in student evaluations in engineering but rather to outline the issues and concerns that engineering academics in a range of countries around the world have dealt with, to give examples of the range of research activities undertaken to investigate the effectiveness of collecting particular types of feedback or the impact of actions following feedback and perhaps to offer some suggestions to other engineering academics tackling similar issues. The overarching message coming from the contributors to this book is that student feedback is an important part of reviewing and enhancing the quality of the entire learning environment, however, it needs to have a purpose and should be ongoing.

References

Astin, A.W. (1982) 'Why not try some new ways of measuring quality?' *Educational Record*, 63(2), 10–15.

Bennett, L. and Nair, C.S. (2010) 'A recipe for effective participation rates for web based surveys'. *Assessment and Evaluation Journal*, 35(4), 357–66.

Brown, R. (2004) *Quality Assurance in Higher Education: The UK Experience since 1992*. London: RoutledgeFalmer.

Coates, H. (2006) *Student Engagement in Campus-based and Online Education: University Connections*. London: Taylor and Francis.

Doyle, K.O. (1975) *Student Evaluation of Instruction*. Lexington, MA: D.C. Heath.

Edström, K. (2008) 'Doing course evaluation as if learning matters most'. *Higher Education Research and Development*, 27(2), 95–106.

Feldman, K.A. (1976) 'The superior college teacher from the students' view'. *Research in Higher Education*, 5(3), 243–88.

Fraser, B.J. (1991) 'Two decades of classroom environment research'. In B.J. Fraser and H.J. Walberg (eds.), *Educational Environments: Evaluation, Antecedents and Consequences*, pp. 3–27. London: Pergamon.

Fraser, B.J. (1994) 'Classroom and school climate'. In D. Gable (ed.), *Handbook of Research on Science Teaching and Learning: A Project of the National Science Teachers Association*. New York: Macmillan.

Green, D. (ed.) (1994) *What is Quality in Higher Education?* Buckingham, UK: Society for Research in Higher Education and Open University Press.

Harvey, L. (1998) 'An assessment of past and current approaches to quality in higher education'. *Australian Journal of Education*, 42(3), 237–55.

Harvey, L. (2001) *Student Feedback*. Report for the Higher Education Funding Council for England. Birmingham, UK: Centre for Research into Quality, Birmingham City University. Available online at: *http://www0.bcu.ac.uk/crq/publications/studentfeedback.pdf* (accessed March 2011).

Kember, D., Leung, D.Y.P. and Kwan, K.P. (2002) 'Does the use of student feedback questionnaires improve the overall quality of teaching?' *Assessment and Evaluation in Higher Education*, 27(5), 411–25.

Kogan, M. (1990) 'Fitting evaluation within the governance of education'. In M. Granheim, M. Kogan and U.P. Lundgren (eds.), *Evaluation as Policy Making: Introducing Evaluation into a National Decentralised Educational System*. London: Jessica Kingsley Publishing.

Marsh, H.W. (1984) 'Students' evaluations of university teaching: dimensionality, reliability, validity, potential biases and utility'. *Journal of Educational Psychology*, 76(5), 707–54.

Marsh, H.W., Overall, J.U. and Kesler, S.P. (1979) 'Class size, students' evaluations, and instructional effectiveness'. *American Educational Research Journal*, 16(1), 57–70.

Nair, C.S. (2011) 'Students' feedback an imperative to enhance quality in engineering education'. *International Journal of Quality Assurance in Engineering and Technology Education*, 1(1), 58–66.

Nair, C.S. and Mertova, P. (2011) *Student Feedback: The Cornerstone to an Effective Quality Assurance System*. Cambridge, UK: Chandos.

Nair, C.S., Patil, A. and Mertova, M. (2009) 'Re-engineering graduate skills – a case study'. *European Journal of Engineering Education*, 34(2), 131–9.

Patil, A.S. (2005) 'The global engineering criteria for the development of a global engineering profession'. *World Transaction on Engineering Education*, 4(1), 49–52.

Radcliffe, D.F. (2005) 'Innovation as a meta attribute for graduate engineers'. *International Journal of Engineering Education*, 21(2), 194–9.

Richardson, J.T.E. (2005) 'Instruments for obtaining student feedback: a review of the literature'. *Assessment and Evaluation in Higher Education*, 30(4), 387–415.

Williams, J. and Cappuccini-Ansfield, G. (2007) 'Fitness for purpose? National and institutional approaches to publicising the student voice'. *Quality in Higher Education*, 13(2), 159–72.

Index

Printed and bound by CPI Group (UK) Ltd, Croydon, CR0 4YY

03/10/2024

01040437-0019